CITES and Timber

A guide to CITES-listed tree species

Madeleine Groves
Catherine Rutherford

Kew Publishing
Royal Botanic Gardens, Kew

First published in 2015 by
Royal Botanic Gardens, Kew
Richmond, Surrey, TW9 3AB, UK
www.kew.org

ISBN 978-1-84246-592-9
ISBNe 978-1-84246-593-6

Distributed on behalf of the Royal Botanic Gardens, Kew in North America by the University of Chicago Press, 1427 East 60th Street, Chicago, IL 60637, USA.

British Library Cataloguing in Publication Data
A catalogue record for this book is available from the British Library.

Design and page layout: Christine Beard
Project editor: Catherine Rutherford

Printed in the UK by Blissetts
The paper used for this publication is FSC certified

CONTENTS

INTRODUCTION

The aim of this guide is to introduce you to the tree species that are regulated under the Convention on International Trade in Endangered Species of wild fauna and flora – CITES – and to provide guidance on the key issues regarding the implementation of the Convention for this important group of plants. The guide does not cover all of the woody and tree species regulated under CITES, rather it concentrates on those species found in significant trade for their timber and parts and derivatives or those newly listed in CITES. Subjects covered in this guide include where to find information about the CITES listing; which parts and derivatives are in trade and whether they are regulated or not; identification techniques available; and key resources on where to find more assistance or information. This publication is also available digitally (see *KEY RESOURCES* at the end of the guide).

REGULATING THE TIMBER TRADE

There are a range of national, regional and international guidelines, legislation and treaties in place to verify, monitor and regulate the global harvesting and trade in timber to ensure it is legal, sustainable and traceable. They include promoting good forest governance; independent forest monitoring; chain of custody certification; legality and due diligence verification when procuring, selling and processing timber; and the regulation of the trade in illegal and unsustainably logged timber.

CITES

This international treaty came into force in 1975 (**http://www.cites.org/eng**) and regulates the international trade in plants and animals threatened through trade, listing them in one of three Appendices (I, II and III). In order to implement the Convention each Party has to establish a Management Authority and at least one Scientific Authority, and trade is regulated by means of a permit system. When trade threatens a species' survival in the wild it can be proposed for listing by one or more of the Parties to CITES, either at the Conference of the Parties (CoP) or, if the Party is a range state, unilaterally at any time on Appendix III. Whereas for CITES Appendix I and II listed species checks are carried out to ensure sustainability and that the specimen was not obtained in contravention of national laws, only the contravention checks are required for Appendix III species. The Convention does allow for Parties to take "stricter measures", such as imposing stricter export or import requirements.

Recommendations, guidance and interpretation of the Convention text are provided through CITES Decisions, Resolutions and Notifications. Those currently in effect are found under *DOCUMENTS* on the CITES home website (**http://www.cites.org/eng**). Current CITES Resolutions relating to timber are as follows:

RESOLUTION	SCOPE
Resolution Conf. 10.13 (Rev. CoP 15) Implementation of the Convention for timber species **http://www.cites.org/eng/res/10/10-13R15.php**	Provides guidance and information on classifying monospecific tree plantations as artificially propagated (as defined in Resolution Conf. 11.11 (Rev. CoP15)); HS classification codes of the Harmonized System of the World Customs Organization for timber products (e.g. logs and sawn wood); steps to take when amending a CITES timber listing.
Resolution Conf. 11.11 (Rev. CoP15) Regulation of the trade in plants **http://www.cites.org/eng/res/11/11-11R15.php**	Contains the CITES definition of artificial propagation. Key criteria are that artificially propagated Appendix I / Annex A species can be treated as Appendix II / Annex B species allowing trade for commercial purposes; and that cultivated parental stock must have been *"established in accordance with the provisions of CITES and relevant national laws"*.
Resolution Conf. 13.6 (Rev. CoP16) Implementation of Article VII, paragraph 2, concerning 'pre-Convention' specimens **http://www.cites.org/eng/res/13/13-06R16.php**	There is an exemption from the Convention rules for "pre-Convention" specimens. A guide to what this term means and the differences between how CITES and the EU Wildlife Trade Regulations (WTR) interpret this exception can be found at **http://ec.europa.eu/environment/cites/pdf/referenceguide_en.pdf**
Resolution Conf. 16.10 Implementing the Convention for agarwood producing taxa **http://www.cites.org/eng/res/16/16-10.php**	Details on how artificial propagation, management and trade control, and non-detriment findings relate to agarwood taxa (*Aquilaria* and *Gyrinops* species).
Resolution Conf. 16.8 Frequent cross-border non-commercial movements of musical instruments **http://cites.org/eng/res/16/16-08.php**	To facilitate the non-commercial movement of musical instruments made from CITES-listed species a certificate ("passport") may be issued for frequent cross-border, non-commercial movement of a musical instrument. This includes, but is not restricted to, personal use, performance, display and competition.

European Union (EU) – Implementation of CITES

All member states of the EU are Parties to CITES. Due to the European Single Market and lack of internal border controls they implement CITES uniformly through a set of regulations called the EU Wildlife Trade Regulations (EU WTR – **http://ec.europa.eu/environment/cites/home_en.htm**). The species covered by these Regulations are listed in four Annexes (A, B, C and D). The provisions of these Regulations go beyond those of CITES, including the requirement for an import permit for Appendix II / Annex B species, stricter requirements for the import of pre-Convention material and the possibility to impose import restrictions for certain species/country combinations. A Reference Guide to the EU WTR can be found at **http://ec.europa.eu/environment/cites/pdf/referenceguide_en.pdf** and a guide on the differences between CITES and the EU WTR is found at **http://ec.europa.eu/environment/cites/pdf/differences_b_eu_and_cites.pdf**

The Regulations establish three main groups: the **Committee on Trade in Wild Fauna and Flora**, the **Scientific Review Group (SRG)** and the **Enforcement Group**. Where a member state's CITES Scientific Authority has doubts over the sustainability of an import into the EU this may lead to other member states refusing similar imports and the application being discussed at a meeting of the SRG, which are held four times a year. The SRG may form one of three opinions (**http://ec.europa.eu/environment/cites/pdf/srg/def_srg_opinions.pdf**) – positive, negative or a no opinion (the latter with three possible options). These may lead to EU import suspensions and ultimately a restriction being put in place for that species/country combination. The SRG opinions arising from a meeting are posted online (within five days) on Species + (**www.speciesplus.net**).

For links to a range of EU national agencies concerned with CITES and wildlife trade go to **http://ec.europa.eu/environment/cites/links_national_en.htm**

Forest Law Enforcement, Governance and Trade (FLEGT)

Through its FLEGT Action Plan, published in 2003, the EU initiated a number of regulations and measures to tackle illegal logging and promote legal and sustainable forestry (**http://www.euflegt.efi.int/home**). The two main FLEGT initiatives are the FLEGT licensing scheme, negotiated under bilateral Voluntary Partnership Agreements (VPAs) with partner timber-producing countries, and the implementation of an EU Timber Regulation (EUTR). Under the FLEGT agreement the VPA partner country issues a FLEGT licence for specified timber products. The licence is the proof of legality (it is not required for Annex A, B or C listed species). It accompanies the shipment and must be verified by the EU member state's appointed competent authority (nominated to enforce the FLEGT regulation) before the shipment is allowed entry into to the EU.

EU Timber Regulation (EUTR)

The EU Timber Regulation (**http://ec.europa.eu/environment/forests/timber_regulation.htm**) came into force on 3rd March 2013 to combat the placing on the EU market of illegal timber and products derived from them (as laid out in the Annex to the EUTR **http://eur-lex.europa.eu/LexUriServ/LexUriServ.do?uri=OJ:L:2010:295:0023:0034:EN:PDF**). It is applicable to those who first place timber on the EU market ("Operators") with an emphasis that they exercise "due diligence" and for those further down the supply chain ("Traders") to keep records of suppliers. Timber and timber products covered by valid FLEGT or Wildlife Trade Regulations (Annex A, B, C specimens) are considered to comply with the permit requirements of this Regulation.

The Lacey Act

This U.S. Act states it is illegal to import, export, sell, acquire, or purchase fish, wildlife or plants that are taken, possessed, transported, or sold in violation of U.S. or Indian law or in interstate or foreign commerce involving any fish, wildlife, or plants taken, possessed or sold in violation of a State or foreign law. The Lacey Act covers CITES species and in 2008 it was amended to include products made from illegally logged woods (**http://www.fws.gov/international/laws-treaties-agreements/us-conservation-laws/lacey-act.html**).

Illegal Logging Prohibition Act (Australian Government 2012)

This Act prohibits the importation and processing of illegally logged timbers, with importers and processors required to make a declaration to customs at time of import to show compliance with the due diligence requirements under the Act (**http://w ww.timberduediligence.com.au/**).

UNDERSTANDING A CITES LISTING

When a species is listed in the CITES Appendices or EU Annexes it is important to understand the information attached to the listing to ensure it is implemented and enforced correctly. There are numerous sites to help you including the CITES website (**http://www.cites.org**) and the EU website on CITES implementation (**http://ec.europa.eu/environment/cites/legislation_en.htm**). One central point of information is called **Species +** (**www.speciesplus.net**), a website developed by UNEP-WCMC (**http://www.unep-wcmc.org**) and the CITES Secretariat to assist Parties with implementing a number of multilateral environmental agreements (MEAs) including CITES. Below are a series of basic questions that may help you understand the scope of a timber listing.

1. *Is the specimen listed in the CITES Appendices and/or the EU Annexes?*

For a full list of CITES regulated species check the current Appendices (**http://www.cites.org/eng/app/appendices.php**), EU Annexes (http://eur-lex.europa.eu/legal-content/EN/TXT/?uri=CELEX:32014R1320) or Species + where you can click on *DOWNLOAD SPECIES LISTS* or type in the name as shown in the example below.

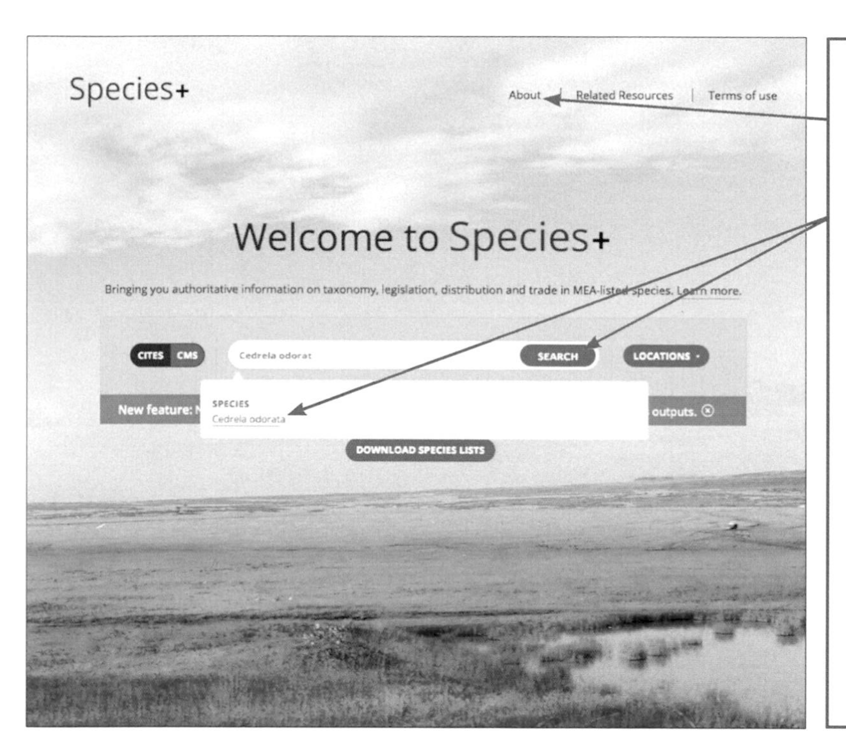

Species + is set to **CITES default.** For guidance on how to use Species + click the *ABOUT* button

Check to see if a species is listed by typing in the name (scientific or common) and clicking *SEARCH*. As you are typing Species + may suggest a name to choose from.

Where possible use the scientific name. Using common names can be confusing as more than one species or genera may be assigned to that name.

If in doubt, check with the trader and your Scientific Authority to clarify what they understand to be the accepted scientific name. Alternatively, check to see if a CITES accepted standard reference or checklist is available for that species / genus (**http://www.cites.org/eng/resources/publications.php**).

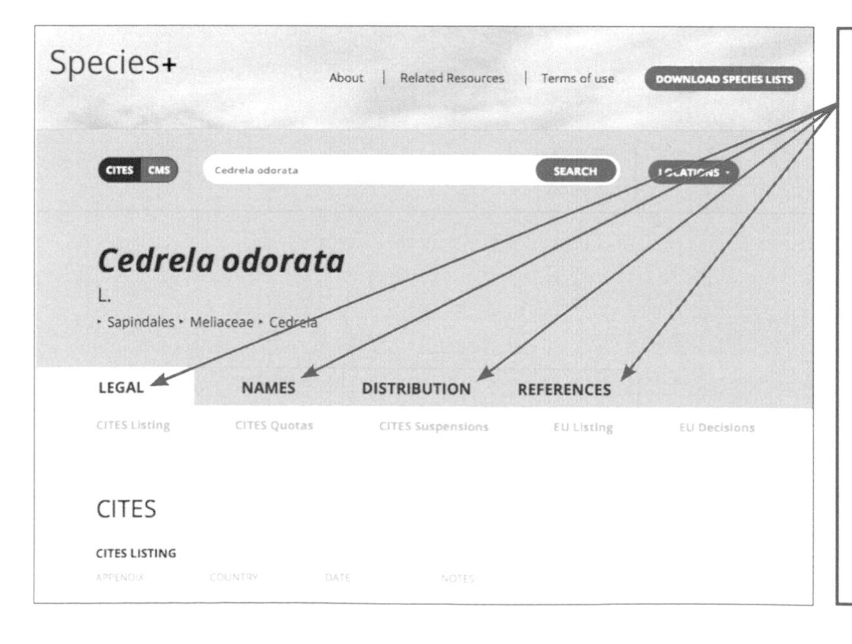

If CITES-listed, an information page with four tabs will appear – *LEGAL* (information on the CITES listing/suspensions/quotas/reservations and EU listing/suspensions/opinions); *NAMES* (scientific and common names); *DISTRIBUTION* (countries and territories); and *REFERENCES* (for distribution and CITES standard references)

If not CITES-listed the website will state there are no results for that species. The reasons for this may include there being a delay in the listing coming into force or it being published in the EU Annexes, or you may have misspelled the name. Check the outcome of the last CoP or CITES Notifications for more information **http://www.cites.org/eng/notif/2014.php**

2. *Which parts and derivatives are regulated?*

When a species is listed under CITES it is necessary to understand whether all parts and derivatives, live or dead, are regulated or not. To clarify this listings may be accompanied by notes or text called 'annotations'. Annotations for plant listings are often complex or require an understanding of the terms used in them. It helps to check the annotation and any definitions with other Scientific Authorities, in the Interpretation section of the Appendices or EU Annexes and in the CITES *GLOSSARY* (*RESOURCES-TERMINOLOGY* in the CITES website (**www.cites.org/eng/resources/terms/index.php**). The Glossary contains definitions of some of the terms used in annotations and are meant as general guidance only as they may not be accepted by all CITES Parties. The CITES Standing Committee, together with the CITES Plants and Animal Committees, is currently working to ensure that the annotations, particularly for plants, are easier to understand and implement.

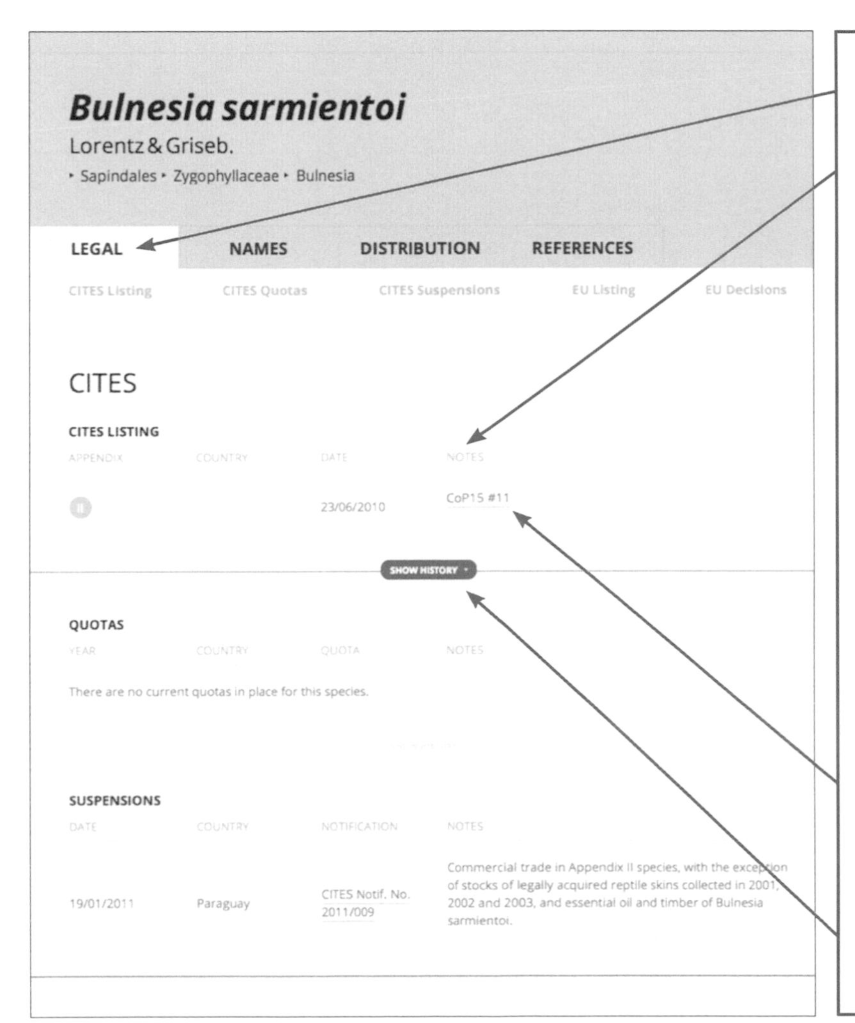

Scroll down the *LEGAL* tab to find the *CITES LISTING* section. The *EU LISTING* section is at the end of the *LEGAL* tab. For both sections you will find the annotation under the *NOTES* column. Click on the annotation and a text box appears outlining the scope of the annotation.

Appendix I / Annex A listed plants, apart from orchids, do not have annotations. Therefore all parts and derivatives, including live or dead specimens, are regulated.

For Appendix II and III / Annex B and C listed plants, if no annotation is present, then all parts and derivatives, live or dead, are regulated. If an annotation is present it will define what is regulated or not. For Annex D species only live specimens are regulated unless the listing is annotated otherwise to denote whether specific parts and derivatives are regulated.

In this example the annotation for *Bulnesia* is #11 meaning only logs, sawn wood, veneer sheets, powders and extracts are regulated. CoP15 denotes the last CoP which had a bearing on the listing and annotation.

Click on *SHOW HISTORY* to see the history of the listing and annotation from the first date the species was listed.

Annotations – there are a number of different types of annotations for plants:

- **# numbers e.g. #5, #11, #12** – these define the parts and derivatives that are subject to the provisions of the Convention
- **Small footnote numbers e.g [11]** – these provide information on the listing, such as any special conditions related to the listing e.g. how specimens must be shipped, whether cultivated varieties are regulated or not
- **Additional text** – this is usually in parentheses and located in the Appendices / Annexes themselves, and provides information on taxonomic issues relevant to the listing or defines which populations are regulated e.g. "populations of..."

3. *Are all populations or only certain populations regulated?*

It is important to understand whether a species / genus listing covers all populations or only certain populations (e.g. populations of Madagascar, populations of the Neotropics) as this will determine whether permits are required, and if so which ones. For information on permit requirements check the CITES website and the EU WTR Reference Guide (under *KEY RESOURCES*) or for more formal guidance contact the Management Authority of the country of export and import, the CITES Secretariat or the EU Commission.

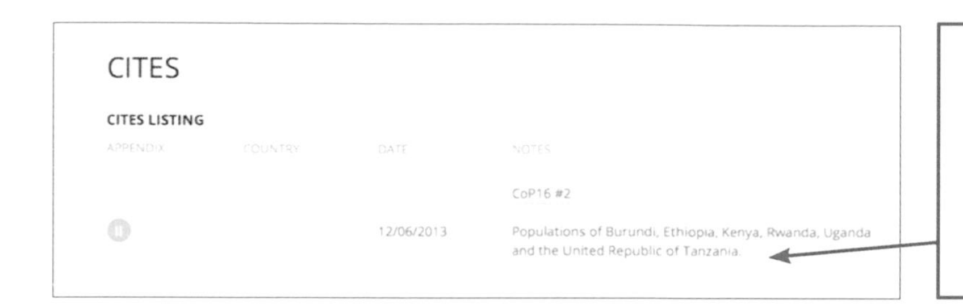

For Appendix II / Annex B species the Species + database will highlight whether specific populations are listed under the *NOTES* heading. In this example (*Osyris lanceolata*) only regulated specimens from the Parties outlined under the NOTES heading are regulated.

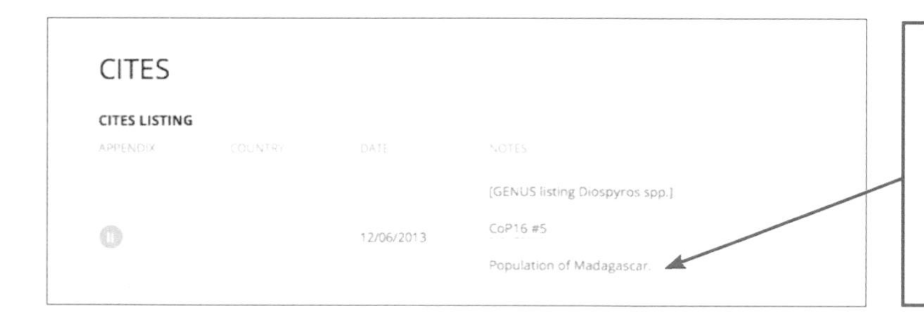

In this example *Diospyros* spp. (all species) are listed in Appendix II / Annex B with a #5 annotation. However, the "Populations of Madagascar" text restricts the scope of the listing to only cover regulated specimens from species endemic to Madagascar.

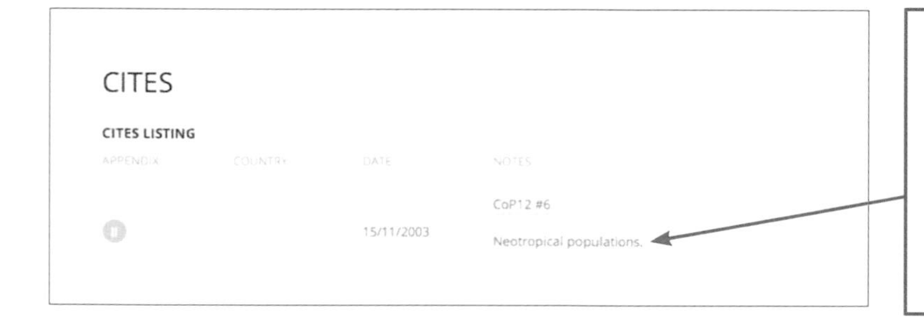

For *Swietenia macrophylla* the Appendix II / Annex B listing only covers the populations that fall within the "Neotropics" (i.e. Central and South America and the Caribbean). If, for example, timber from this species originated from a plantation in Fiji, which is outside of the "Neotropics", it would not be regulated.

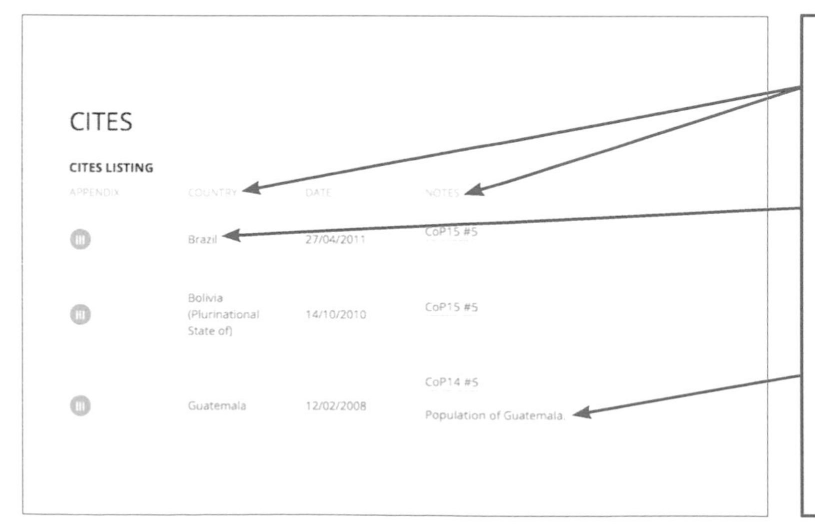

For Appendix III / Annex C species, look at the *COUNTRY* and *NOTES* headings (under the *CITES LISTING* or *EU LISTING* tabs) to check if all or only certain populations are listed.

If a country name is only found under the *COUNTRY* heading all populations were listed by that country (illustrated here by Brazil and Bolivia).

You may find the same country name under both the *COUNTRY* and *NOTES* headings. However, under the *NOTES* heading the name is preceded by "populations of…". This means that the country only listed its own populations (illustrated in this example by "populations of Guatemala").

4. *Are CITES export quotas in place?*

CITES export quotas are established by Parties unilaterally or are set by the Conference of the Parties as a tool to help regulate trade. They usually run annually (1st January to 31st December) but harvesting periods do not always follow this timeframe. Changes or updates are relayed through a Notification to the Parties. Export quotas since 2000 are available on the CITES website (**http://www.cites.org/eng/resources/quotas/index.php**)

Scroll down through the *LEGAL* tab until you reach the *QUOTA* section. Here you will find the current annual quota per country together with the products it applies to and the quota level.

In this example, the only country with a 2015 quota for *Prunus africana* is Cameroon. The quota is for 974,853 kg of dry bark.

Click on *SHOW HISTORY* for previous quotas, which products they applied to and the previous quota level.

Check Species + for export quota updates or the CITES website for a list of annual export quotas (**http://cites.org/eng/resources/quotas/index.php**).

5. *Has a country taken out a reservation on the listing?*

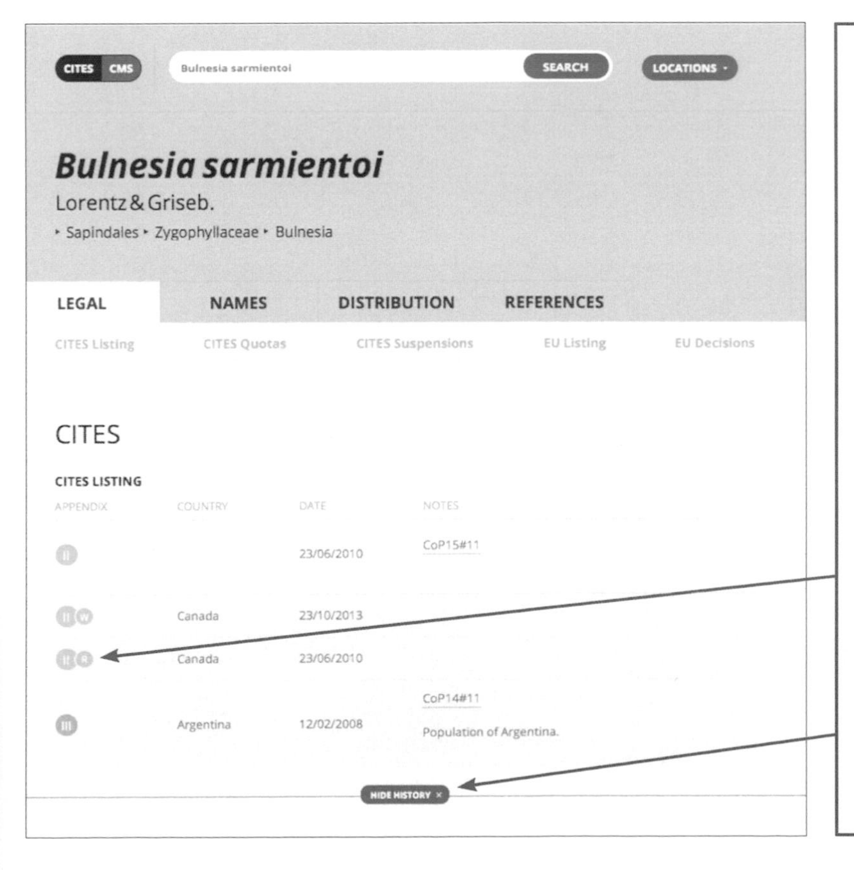

A reservation is a unilateral statement by a Party not to be bound by the CITES provisions for a species listing. A Party may take out a reservation if a species is not yet covered by its own national legislation.

That Party is treated as a non-Party with regard to that species and if it exports to a non-Party or another Party with the same reservation, no CITES documents are required. An export permit is required if the Party with a reservation exports to a Party that has not taken out a reservation on the same species.

Reservations, if any, will be listed under the *CITES LISTING* section in the *LEGAL* tab. In this example the "R" means Canada took out a reservation on the Appendix II listing of *Bulnesia sarmientoi* from 23/06/2010 until 23/10/2013 when it was withdrawn ("W").

Click on *SHOW HISTORY* to see previous reservations.

6. *Are there specific CITES international trade suspensions or EU opinions in place for this listing?*

Species + notes any national export suspensions or CITES international suspensions that affect trade. The CITES Parties may take "stricter measures" than the Convention. As such they may impose stricter import or export conditions. The EU member states, through its SRG, issue opinions relating to stricter conditions for importing material into the EU. They are positive (import is allowed subject to permits being issued), negative (no imports until further communication with the country in question) or no opinion. The latter can take three separate forms, including the possibility that all import import applications must be referred to the SRG for consideration.

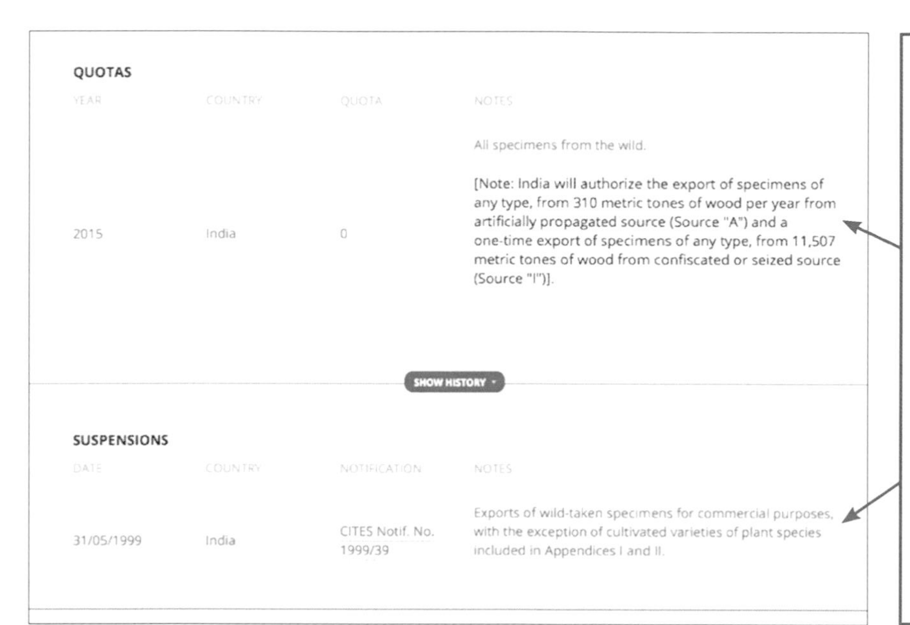

This example uses *Pterocarpus santalinus* (red sandalwood). Scroll down the *LEGAL* tab to the *SUSPENSION* heading for details on current CITES and national suspensions.

India is implementing a zero export quota for all wild specimens but allows the export of seized material.

India also has a national suspension on the export of wild Appendix II / Annex B specimens, such as red sandalwood. This suspension came into force on 31/05/1999 and a Notification is given which provides more details (**http://cites.org/sites/default/files/eng/notif/1999/039.shtml**).

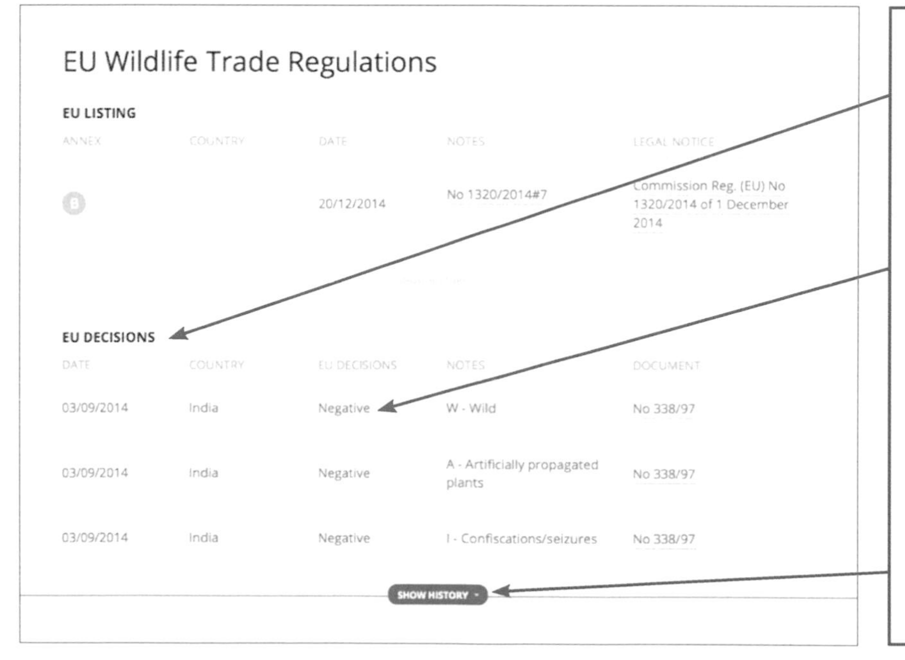

Current EU decisions and opinions are found under the *EU DECISIONS* heading in the *LEGAL* tab. Details of what they apply to and the date they came into force are provided.

The EU has put in place a negative opinion (no EU imports allowed) for all wild, artificially propagated and confiscated/seized material of red sandalwood from India. If new information is provided to the SRG this opinion may be overturned at its next meeting. If not, the opinion may remain in place.

Click on the *SHOW HISTORY* to see all previous EU opinions.

SPECIES PAGES

The following information is provided for each species / genus:

Distribution – covers the countries / territories that the species / genus are native to.*

Uses – the major uses for the taxa and products in trade (regulated or not).

Trade – the dataset used to inform this section has been taken from the CITES Trade Database (**http://trade.cites.org/**) for the years 2003–2013.

Plantations / artificial propagation – highlights whether a species / genus is grown in plantations and / or is artificially propagated and whether timber or timber products from these sources are in trade.

CITES international trade suspensions, export quotas and reservations – current CITES international trade suspensions, export quotas and reservations are noted.

EU Decisions – this section outline those trade suspensions and opinions currently put in place by the European Union. These may be stricter than CITES.

Scientific and common names – CITES adopts "Standard References" to be used by Parties when referring to the scientific names of taxa listed on the Convention. These references provide the basis on which names should be used on CITES permits and annual reports – the accepted names – as adopted by a CoP. The references, where possible, include the major synonyms – the non-accepted names – that apply to these taxa. The list of Standard References is updated at every CoP based on the recommendations of the CITES Animal and Plants Committees. The list of "Standard References" is included in a CITES Resolution (for flora see Annex 2 of Resolution Conf. 12. 11 (Rev. CoP16) Standard Nomenclature: **http://www.cites.org/eng/res/12/12-11R16.php**). Where a standard reference has been adopted for a species or genus this reference is given. Where no standard reference is available we recommend a suitable source, but this is merely our recommendation. In such cases you can seek clarification or a definitive decision from the Scientific Services Unit (**http://www.cites.org/eng/disc/sec/staff.php**) of the CITES Secretariat who may confer with the Nomenclature Specialist of the CITES Plants Committee to provide you with more formal guidance on the appropriate scientific name.

Details on the CITES listing – the date the species / genus were first listed is provided along with the date of the current listing and annotation. For EU listings the current listing and annotation date applies to the latest version of the Commission Regulation amending the list of species regulated under Council Regulation (EC) No 338/97.

Product pictures – pictures have been provided of products seen in international trade. As space is a limiting factor in the guide the coverage is not comprehensive. Products regulated by CITES have a **RED** tab and those not regulated have a **GREEN** tab. The photographs used are not, in all cases, true representations of the species in trade, rather they are meant as a guide to the products in trade.

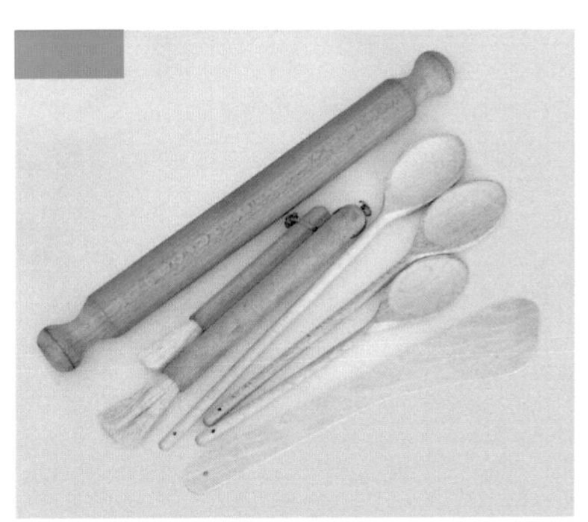

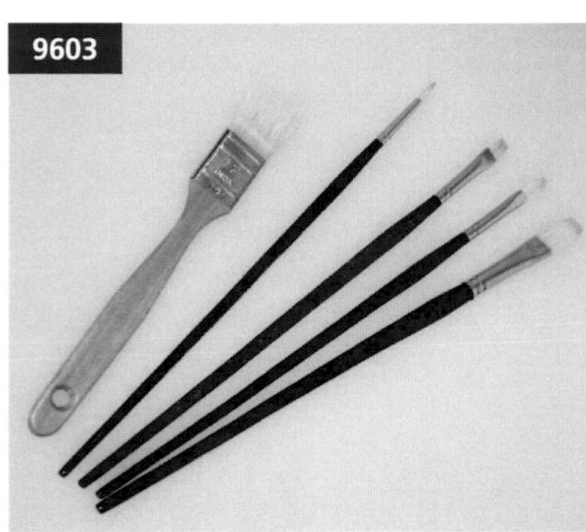

Tariff codes – Commodity tariff chapter codes are provided against each picture of regulated products (red tab). A guide to tariffs codes can be found in the current edition of the International Integrated Tariffs. Due to the continual updating of specific codes, tariff CHAPTER CODES are listed rather than full individual tariff codes.

Picture credits / copyright – these are provided on page 91.

* Country abbreviations include Taiwan for Tawain (Province of China); UK (United Kingdom); USA (United States of America); EU (European Union).

Abies guatemalensis
Guatemalan Fir

Distribution

This is the only species in the genus *Abies* currently listed under CITES. The species is native to El Salvador, Guatemala, Honduras and Mexico.

Uses

The timber has historically been used for construction purposes, and in the manufacture of tools, roof shingles, wood carvings and charcoal for domestic markets. Illegal harvesting for the domestic Christmas tree and Christmas decoration market is threatening the species.

Trade

This species is listed in Appendix I / Annex A and international trade in wild-sourced specimens for commercial purposes is prohibited. Commercial trade in artificially propagated specimens is permitted. Domestic use of this species accounts for the majority trade. The CITES Trade Database indicates little or no international trade with a small number of wild-collected specimens (fruit, live, seeds) exported from Guatemala and Mexico to the EU (Denmark) for scientific purposes.

As an Appendix I / Annex A listed species there is no annotation therefore all parts and derivatives, live or dead, are regulated.

Plantations/artificial propagation

There are commercial plantations of this species in Guatemala and El Salvador, primarily supplying domestic markets.

CITES international trade suspensions, export quotas and reservations

There are no current CITES international trade suspensions, export quotas or reservations in place for this species.

EU Decisions

There are no current EU suspensions or opinions for this species.

See Species + for details **http://speciesplus.net/#/taxon_concepts/17704/legal**

Plantation

Roof shingles

SCIENTIFIC AND COMMON NAMES	DATE OF LISTING	CURRENT LISTING AND ANNOTATION
Scientific name and author: *Abies guatemalensis* Rehder	**Appendix I** 01/07/1975 **Annex A** 01/06/1997	**Appendix I** 01/07/1975 **Annex A** 20/12/2014 **Annotation** As an Appendix I /Annex A species all parts and derivatives, live or dead, are regulated
Family: Pinaceae		
Common names: English: Guatemalan fir, Mexican fir French: sapin du Guatemala German: Guatemala-tanne Italian: abete del Guatemala Spanish: abeto Mexicano, pinabete, abeto, pashaque, romerillo		
CITES Standard Reference: The CoP has adopted a standard reference for generic names (*The Plant Book*, 2nd Edition, D. J. Mabberley, 1997, Cambridge University Press reprinted with corrections 1998). The Plant List is a useful source of information on scientific names and synonyms for this species (**http://www.theplantlist.org/tpl1.1/record/kew-2609883**). If you require more formal guidance contact the CITES Secretariat.		

Christmas decoration

Tool handles

Aniba rosaeodora
Rosewood, Pau rosa

Distribution

This is one of 48 species in the genus *Aniba* and the only one listed under CITES. It is a large tree native to the tropical rainforests of Brazil, Colombia, Ecuador, French Guiana, Guyana, Peru, Suriname and Venezuela.

Uses

The timber is harvested and distilled to extract the fragrant rosewood oil which is used in high end perfumery, to produce pure essential oils for aromatherapy or to add scent to cosmetics and toiletries, such as soap. The oil may be diluted to produce less expensive fragrances and oils. Oil is also extracted from leaves and smaller branches and research is currently underway to determine if this oil has different olfactory characteristics and quality to that distilled from the heartwood.

Trade

The main product in international trade is rosewood oil. The sole exporter supplying this small scale but valuable trade is Brazil. The USA is the principal importer of the oil, followed by Japan and the EU (France and Germany). The USA is also the largest re-exporter of oil, after the addition of synthetic linalool, followed by France. Current Brazilian national legislation prohibits the domestic and international trade of all native species (CITES and non-CITES species) with an IUCN Red List assessment of critically endangered (CR) and endangered (EN). All trade from Brazil should therefore only be in artificially propagated specimens.

The #12 annotation for this species means only logs, sawn wood, veneer sheets, plywood and extracts are regulated. Definitions of the timber terms are found in the CITES Glossary (**http://www.cites.org/eng/resources/terms/glossary.php**) and in Resolution Conf. 10.13 (Rev. CoP15) (**http://www.cites.org/eng/res/10/10-13R15.php**). A definition of the term extract can be found in the Interpretation section of the Appendices and EU Annexes and in the CITES Glossary as meaning "Any substance obtained directly from plant material by physical or chemical means regardless of the manufacturing process. An extract may be solid (e.g. crystals, resin, fine or coarse particles), semi-solid (e.g. gums, waxes) or liquid (e.g. solutions, tinctures, oil and essential oils)". As stated in the annotation, finished products containing such extracts as ingredients, such as perfumes, are not covered by this annotation and are therefore not regulated.

Rosewood plantation — Artificially propagated seedlings — Logs

Plantations / artificial propagation

There is currently only one authorised commercial plantation for this species, in Maués, Brazil, that is producing oil for export. However, international cosmetic manufacturers are establishing smaller trial plantations, in Peru for example, to ensure tighter control over the sustainability of the ingredients for their products.

CITES international trade suspensions, export quotas and reservations

There are no current CITES international trade suspensions, export quotas or reservations in place for this species.

EU Decisions

There are no current EU suspensions or opinions for this species.

See Species + for details **http://speciesplus.net/#/taxon_concepts/19724/legal**

SCIENTIFIC AND COMMON NAMES	DATE OF LISTING	CURRENT LISTING AND ANNOTATION
Scientific name and author: *Aniba rosaeodora* Ducke	**Appendix II** 23/06/2010 **Annex B** 15/08/2010	**Appendix II** 12/06/2013 **Annex B** 20/12/2014 **Annotation** #12 Logs, sawn wood, veneer sheets, plywood and extracts. Finished products containing such extracts as ingredients, including fragrances, are not covered by this annotation
Family: Lauraceae		
Common names: **English:** Brazilian rosewood, rosewood, rosewood oil **French:** bois de rose, bois-de-rose-femell, carcara **German:** rosenholzbaum **Italian:** legno di rose **Portuguese:** pau rosa **Spanish:** palo de rosa, palo de rose **Swedish:** doftrosendträd		
CITES Standard Reference: The CoP has adopted a standard reference for generic names (*The Plant Book*, 2nd Edition, D. J. Mabberley, 1997, Cambridge University Press reprinted with corrections 1998). The Plant List is a useful source of information on scientific names and synonyms for this species (**http://www.theplantlist.org/tpl1.1/record/kew-2639744**). If you require more formal guidance contact the CITES Secretariat.		

Oil distillation

Essential oil – finished product

Unfiltered oil

Aquilaria and *Gyrinops*
Agarwood

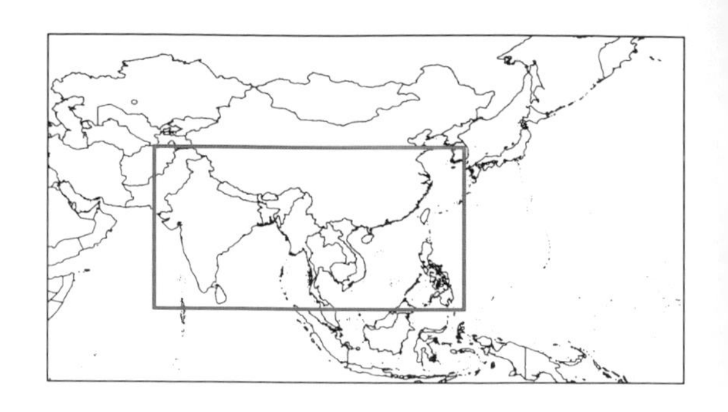

Distribution

The main agarwood producing tree species are in the genera *Aquilaria* and *Gyrinops*. All species in these two genera are listed under CITES. There are some 20 species of *Aquilaria* native to Bangladesh, Bhutan, Cambodia, China, India, Indonesia, Lao People's Democratic Republic, Malaysia, Myanmar, Papua New Guinea, Philippines, Singapore, Thailand and Viet Nam. There are some nine *Gyrinops* species which are native to India, Indonesia and Sri Lanka. The genus *Gonystylus* (ramin) has also been linked to the agarwood trade but is more commonly traded as timber and timber products. See Species + for individual species distribution:

Aquilaria species –
http://speciesplus.net/#/taxon_concepts?taxonomy=cites_eu&taxon_concept_query=Aquilaria&geo_entities_ids=&geo_entity_scope=cites&page=1

Gyrinops species –
http://speciesplus.net/#/taxon_concepts?taxonomy=cites_eu&taxon_concept_query=Gyrinops&geo_entities_ids=&geo_entity_scope=cites&page=1

Uses

The infection of *Aquilaria* and *Gyrinops* trees by fungal pathogens produces a resinous heartwood that is highly prized for its fragrant properties. Oil is distilled from the wood, powder or sawdust and used in the fragrance and cosmetics / toiletries industries. Other products in trade include wood chips of various sizes and powder, sold both prior to the distillation process (non-exhausted) and post the distillation process (exhausted). Exhausted powder may then be further processed and compressed into statues, incense cones or sticks. The wood is made into carvings, beads (for prayer or decoration) or traded as highly prized items. Agarwood is also an ingredient in traditional and patented medicines.

Trade

Agarwood products are found in international trade as crude, semi-finished or finished products from both wild and artificially propagated sources. Processing of agarwood products may occur in countries of export or in the major re-exporting Parties, in particular the Middle East. For *Aquilaria* species wood chips and powders (both non-exhausted and exhausted) are the major products in trade followed by pieces of wood /sawn wood, oils and extracts. The major exporters are Malaysia, Indonesia and Thailand. The major importers are Saudi Arabia, United Arab Emirates (UAE), Kuwait and Europe (France, Germany, Italy and the UK). The major re-exporting countries are Saudi Arabia, Kuwait and UAE. The

Artificially propagated seedlings

Resinous heartwood

Wood chips – unfinished product

trade in *Gyrinops* species is much smaller than that of *Aquilaria* and the main product in trade is wood chips. The major exporter of *Gyrinops* products is Indonesia, imported by Saudi Arabia, Singapore and UAE. The major re-exporters of this genus are Saudi Arabia and Singapore.

Trade in wild-sourced agarwood products is not permitted by most range states (page 11, point 6 **http://www.cites.org/sites/default/files/common/com/pc/20/inf_docs/E20-07i.pdf**). However, trade in products from artificially propagated sources is allowed subject to permits being issued.

Note the following issues regarding the #14 annotation:

- Finished products packaged and ready for retail trade – the annotation states that products traded in this state are not regulated. A definition of this wording can be found in the Interpretation section of the Appendices and Annexes and in the CITES Glossary (**http://www.cites.org/eng/resources/terms/glossary.php**). This means that if a regulated agarwood product, for example non-exhausted powder, fits the definition of a finished product packaged and ready for retail trade then it is not regulated. This does not apply to beads, prayer beads and carvings.
- Identification of agarwood products in trade – distinguishing between agarwood products in trade may prove problematic e.g. the difference between non-exhausted and exhausted powder (the latter is lighter in colour with little odour).

There is also an exemption under the CITES personal and household effects derogation (Resolution Conf. 13.7 (Rev. CoP16) **http://www.cites.org/eng/res/13/13-07R16.php**) for specimens of agarwood. Up to 1 kg woodchips, 24 ml oil and two sets of beads or prayer beads (or two necklaces or bracelets) per person is permitted without permits (**Note**: personally owned or possessed specimens will be exempted as personal effects if both the countries of import and export implement the personal and household effects exemption for the species).

Plantations / artificial propagation

There are commercial mono- and mixed plantations of *Aquilaria* and *Gyrinops* as the technology to artificially innoculate trees and produce the resinous heartwood is now available. Trees of both genera are also grown in mixed community farms and gardens within its natural range. Timber or timber products from all of these production systems are in international trade and may be declared as artificially propagated material. Check with the country of origin and Resolutions Conf. 11.11 (Rev. CoP15) (**http://www.cites.org/eng/res/11/11-11R15.php**), Resolution Conf. 10.13 (Rev. CoP15) (**http://www.cites.org/eng/res/10/10-13R15.php**) and Resolution Conf. 16.10 (**http://www.cites.org/eng/res/16/16-10.php**) for guidance on whether the material fits the definition of artificially propagated for agarwood.

CITES international trade suspensions, export quotas and reservations

There are current CITES international trade suspensions for some species of *Aquilaria*.

There are current export quotas in place for *Aquilaria* species. Check Species + for details or check the annual export quotas on the CITES website (**http://www.cites.org/eng/resources/quotas/index.php**).

There are current reservations in place for a number of *Aquilaria* and *Gyrinops* species.

There are current national export bans in place for some species of *Aquilaria* and *Gyrinops*.

Wood chips – finished product

Compressed exhausted powder

Prayer beads

EU decisions

There are a number of current EU opinions in place for both *Aquilaria* and *Gyrinops* species.
See Species + for details:
Aquilaria species – **http://www.speciesplus.net/#/taxon_concepts?taxonomy=cites_eu&taxon_concept_query=aquilaria&geo_entities_ids=&geo_entity_scope=cites&page=1** and
Gyrinops species – **http://www.speciesplus.net/#/taxon_concepts?taxonomy=cites_eu&taxon_concept_query=Gyrinops&geo_entities_ids=&geo_entity_scope=cites&page=1**

SCIENTIFIC AND COMMON NAMES	DATE OF LISTING	CURRENT LISTING AND ANNOTATION
Scientific names and author: *Aquilaria* Lam**:** 20 species *Gyrinops* Gaertn: 9 species	*Aquilaria malaccensis* **Appendix II**: 16/02/1995 **Annex B**: 01/06/1997	*Aquilaria* spp. **Appendix II**: 12/06/2013 **Annex B**: 20/12/2014
Family: Thymelaeceae	*Aquilaria* spp. **Appendix II**: 12/01/2005 **Annex B**: 22/08/2005	*Gyrinops* spp. **Appendix II:** 12/06/2013 **Annex B:** 20/12/2014
Common names: **Chinese:** chen xiang, chénxiāng **English:** agarwood, eaglewood, aloewood, agar, oud, oudh **French:** bois d'aigle **German:** adlerholz **Indonesia:** gaharu **Italian:** agarwood **Laos:** mai ketsana **Malaysia:** a-ga-ru, agur, alim, calambac, gaharu, halim, karas, kareh, kritsanaa, lign-aloes, mai hom **Myanmar:** thit mhwae **PNG:** ghara **Portuguese:** aquilária, madeira de agar, calambuca **Spanish:** madera de Agar **Thai:** mai kritsana **Viet Nam:** trầm hu'o'ng	*Gyrinops* spp. **Appendix II**: 12/01/2005 **Annex B**: 22/08/2005	**Annotation** **#14** All parts and derivatives are regulated, except: – Seeds and pollen – Seedling and tissue cultures obtained in vitro, in solid or liquid media, transported in sterile containers – Fruits – Leaves – Exhausted agarwood powder, including compressed powder in all shapes – Finished products packaged and ready for retail trade, this exemption does not apply to beads, prayer beads and carvings

CITES Standard Reference: The CoP has adopted a standard reference for generic names (*The Plant Book*, 2nd Edition, D. J. Mabberley, 1997, Cambridge University Press reprinted with corrections 1998). The Plant List is a useful source of information on scientific names and synonyms for the species in these genera
(**http://www.theplantlist.org/tpl1.1/search?q=Aquilaria** and **http://www.theplantlist.org/tpl1.1/search?q=Gyrinops**).
If you require more formal guidance contact the CITES Secretariat.

Perfume – finished product

Bracelet

Essential oil – finished product

Araucaria araucana
Monkey Puzzle tree

Distribution

This is the only species in the genus *Araucaria* (19 species) currently listed under CITES. It is a long-lived, tall conifer, commonly known as the monkey puzzle tree or Chilean pine, and is native to the temperate rainforests of southern and central Chile and south west Argentina.

Uses

The monkey puzzle tree has historically been used to make beams in buildings, bridges, piers, pit props, roofs, furniture, ship masts, veneers, plywood, flooring and paper pulp. The pine nuts are edible and are harvested in Chile under strict legislation. This species was introduced into Europe as early as the late 1700s.

Trade

This species is listed in Appendix I /Annex A and the international trade in wild-sourced specimens for commercial purposes is prohibited. Commercial trade in artificially propagated specimens is permitted. The CITES Trade Database indicates that the majority of trade is in artificially propagated live plants and seeds (source code "D" which is artificially propagated material of an Appendix I species). The main exporters of these specimens are Chile and to a much lesser extent Argentina. The main importers are the Netherlands with smaller quantities imported by Switzerland. The main re-exports of live plants are from Chile and the EU (Belgium, Germany, Italy and the Netherlands) to Switzerland. The main re-exports of seed are from New Zealand to the USA. Timber or wood blanks from felled ornamental trees planted outside of its natural range are in international trade for the manufacture of wooden artefacts (e.g. bowls and vases). Fossilised specimens of this species are also for sale on the Internet. As an Appendix I /Annex A listed species there is no annotation therefore all parts and derivatives, live or dead, are regulated.

Plantations / artificial propagation

Commercial plantations are established worldwide, in particular in Chile, Argentina and New Zealand. Artificially propagated seed and specimens of this species are for sale on the Internet and can be found growing in nurseries, parks and gardens, particularly throughout Western Europe, the west and east coast of the USA, New Zealand and south eastern Australia.

Araucaria araucana

Veneer sheet

A. araucana bowl

CITES international trade suspensions, export quotas and reservations

There are no current CITES international trade suspensions, export quotas or reservations in place for this species.

EU Decisions

There are no current EU suspensions or opinions for this species.

See Species + for details **http://www.speciesplus.net/#/taxon_concepts/25968/legal**

SCIENTIFIC AND COMMON NAMES	DATE OF LISTING	CURRENT LISTING AND ANNOTATION
Scientific name: *Araucaria araucana* (Molina) K.Koch	**Appendix II** 01/07/1975	**Appendix I** 13/02/2003
Family: Araucariaceae	**Annex B** 01/06/1997 (Populations of Chile)	**Annex A** 20/12/2014
Common names: **English:** monkey puzzle tree, Chilean pine **French:** désespoir des singes **German:** araukarie, Chilenische araukarie **Spanish:** araucaria, piñonero, pino araucaria, pehuén (from Mapuche), pino Chileno **Swedish:** brödgran		**Annotation** As an Appendix I / Annex A species all parts and derivatives, live or dead, are regulated
CITES Standard Reference: The CoP has adopted a standard reference for generic names (*The Plant Book*, 2nd Edition, D. J. Mabberley, 1997, Cambridge University Press reprinted with corrections 1998). The Plant List is a useful source of information on scientific names and synonyms for this species (**http://www.theplantlist.org/tpl1.1/record/kew-14339**). If you require more formal guidance contact the CITES Secretariat.		

Petrified *Araucaria* cone

A. araucana seeds

Bulnesia sarmientoi
Palo santo, holy wood

Distribution

Bulnesia sarmientoi is the only one of ten species in the genus *Bulnesia* listed under CITES. This species is a deciduous tree known for its aromatic wood. It is native to south-eastern Bolivia through western Paraguay and adjoining sectors of Brazil to northern Argentina, an area called the Gran Chaco (or Región Chaqueña).

Uses

The heavy, hard wood is used in the manufacture of wood flooring, furniture, handicrafts, smoking pipes, mortars, and axes. The wood is distilled to produce an essential oil known as 'guayacol' which is used in the manufacture of perfumes, cosmetics, toiletries and candles. The oil is also mixed with pyrethrum to make mosquito coils. Residual sawdust is treated with solvents to produce 'palo santo' resin used in the manufacture of varnish and dark paints. Small pieces of wood or wood chips, oil and powder compressed into cones or incense sticks are in international trade, in particular over the Internet. Some of these products may be made from non-CITES listed species (e.g. *Bursera graveolens*) which also use the common name 'palo santo'. There are also similarities between *B. sarmientoi* and species in the genus *Guaiacum*, which belongs to the same family as *Bulnesia* (Zygophyllaceae). They both produce oil, the wood has similar characteristics and they share the common names of 'palo santo' and 'guayacán' as well as some of their trade names, such as 'lignum vitae' and 'guaiac'. All *Guaiacum* species are also regulated under CITES.

Trade

Bulnesia sarmientoi is in international trade as timber and semi-finished and finished products. The majority of international trade is in wild-sourced timber exported from Argentina to China. Paraguay exports mainly essential oil for further processing in the fragrance industry, which is imported into China, the USA, the EU (France, Germany, Italy, Spain, Italy, the Netherlands and the UK) and Switzerland. The major re-exporters of oil are the EU (France, Germany and Spain), Switzerland and the USA.

The #11 annotation means only logs, sawn wood, veneer sheets, plywood, powder and extracts are regulated. Definitions of the timber terms are found in the CITES Glossary (**http://www.cites.org/eng/resources/terms/glossary.php**) and in Resolution Conf. 10.13 (Rev. CoP15) (**http://www.cites.org/eng/res/10/10-13R15.php**). Definitions of the terms "extract" and "powder" can be found in the Interpretation to the Appendices and EU Annexes and in the CITES Glossary. The definition of an extract is given as "Any substance obtained directly from plant material by physical or chemical

Logs Logs Utensils

means regardless of the manufacturing process. An extract may be solid (e.g. crystals, resin, fine or coarse particles), semi-solid (e.g. gums, waxes) or liquid (e.g. solutions, tinctures, oil and essential oils)". The definition of a powder is given as "A dry, solid substance in the form of fine or coarse particles".

Plantations/artificial propagation

There are no known commercial plantations or artificial propagation of this species.

CITES trade suspensions, export quotas and reservations

There are no current CITES international trade suspension, export quotas or reservations in place for this species.

There is a current suspension on the export of Appendix II species from Paraguay, but this does not apply to the export and trade in timber and essential oil of *B. sarmientoi*.

EU Decisions

There is a current EU opinion for this species from Paraguay.

See Species + for details **http://speciesplus.net/#/taxon_concepts/23851/legal**

SCIENTIFIC AND COMMON NAMES	DATE OF LISTING	CURRENT LISTING AND ANNOTATION
Scientific name and author: *Bulnesia sarmientoi* Lorentz ex Griseb.	**Appendix III** 12/02/2008 (population of Argentina only) **Annex C** 11/04/2008	**Appendix II** 23/06/2010 **Annex B** 20/12/2014 **Annotation** **#11** Logs, sawn wood, veneer sheets, plywood, powder and extracts.
Family: Zygophyllaceae		
Common names: **English:** holy wood, verawood, lignum vitae, Paraguay lignum vitae, Argentine lignum vitae, guaiac, guaiac wood **French:** bois de gaic **Portuguese:** pau santo **Spanish:** palo santo, guayacán, palo bálsamo		
CITES Standard Reference: The CoP has adopted a standard reference for generic names (*The Plant Book*, 2nd Edition, D. J. Mabberley, 1997, Cambridge University Press reprinted with corrections 1998). The Plant List is a useful source of information on scientific names and synonyms for this species (**http://www.theplantlist.org/tpl1.1/record/kew-2685914**). If you require more formal guidance contact the CITES Secretariat.		

Essential oil — Essential oil — Mosquito coil

Caesalpinia echinata
Pernambuco, pau-brasil

Distribution

Of the 161 species in the genus *Caesalpinia* this is the only species listed under CITES. It is endemic to the eastern region of Brazil in the threatened Atlantic Coastal Rainforest (Mata Atlântica) in the states of Pernambuco, Bahia, Espírito Santo and Rio de Janeiro.

Uses

Historically valued in Europe as a source of red dye, the wood is flexible but durable and was also used for construction purposes. From the late 18th century its popularity led to it being used in the manufacture of violin bows. Currently, pernambuco is the principle wood used to make high quality violin bows.

Trade

The main products in trade are timber and carvings, including wood blanks for the manufacture of violin bows. All exports originate from Brazil. The CITES Trade Database indicates that the majority of the timber is source code "O", meaning it was acquired before the species was listed on CITES (i.e. pre-Convention). The main importers of timber and carvings are China, Switzerland and to a lesser extent the EU (Bulgaria, France, Germany, the Netherlands, and the UK), Japan and the USA. The biggest re-exporters are the USA (to China and EU – France, Germany, Spain, the UK and Bulgaria) and Germany (to China and Switzerland). There are small volumes of wild-sourced timber re-exported from the USA to France, China and Australia. Current Brazilian national legislation prohibits the domestic and international trade of all wild-collected native species (CITES and non-CITES species) with an IUCN Red List assessment of critically endangered (CR) and endangered (EN). All trade from Brazil should therefore only be in artificially propagated specimens.

The #10 annotation means that only logs, sawn wood, veneer sheets, including unfinished wood articles ("blanks") used for the fabrication of bows for stringed musical instruments are regulated. Definitions of the timber terms are found in the CITES Glossary (**http://www.cites.org/eng/resources/terms/glossary.php**) and in Resolution Conf. 10.13 (Rev. CoP15) (**http://www.cites.org/eng/res/10/10-13R15.php**). A finished bow made from *Caesalpinia echinata* is not regulated whereas an unfinished wooden "blank" used to make a bow would require a permit.

Pernambuco tree

Unfinished bow blanks

Stacks of violin blanks

Plantations / artificial propagation

There are no large-scale commercial plantations of this species. Smaller replanting and conservation initiatives, such as those being carried out by the International Pernambuco Conservation Initiative (**http://www.ipci-usa.org/index.html**), are in place in Brazil, but as yet no wood from these sources is in international trade.

CITES international trade suspensions, export quotas and reservations

There are no current CITES international trade suspensions, export quotas or reservations in place for this species.

EU Decisions

There are no current EU suspensions or opinions for this species.

See Species + for details **http://www.speciesplus.net/#/taxon_concepts/18906/legal**

SCIENTIFIC AND COMMON NAMES	DATE OF LISTING	CURRENT LISTING AND ANNOTATION
Scientific name and author: *Caesalpinea echinata* Lam.	**Appendix II** 13/09/2007 **Annex B** 11/04/2008	**Appendix II** 13/09/2007 **Annex B** 20/12/2014 **Annotation** **#10** Logs, sawn wood, veneer sheets, including unfinished wood articles used for the fabrication of bows for stringed musical instruments
Family: Leguminosae		
Common names: English: Brazil wood, pernambuco wood French: bois de pernambouc German: pernambuckholz Portuguese: pau-brasil, ibirapitanga, brasileto, orabutá Spanish: palo brasil, palo pernambuco, palo rosado Swedish: bresilja		
CITES Standard Reference: The CoP has adopted a standard reference for generic names (*The Plant Book*, 2nd Edition, D. J. Mabberley, 1997, Cambridge University Press reprinted with corrections 1998). The Plant List is a useful source of information on scientific names and synonyms for this species (**http://www.theplantlist.org/tpl1.1/record/ild-20094**). If you require more formal guidance contact the CITES Secretariat.		

Unfinished bow blanks

Finished bows

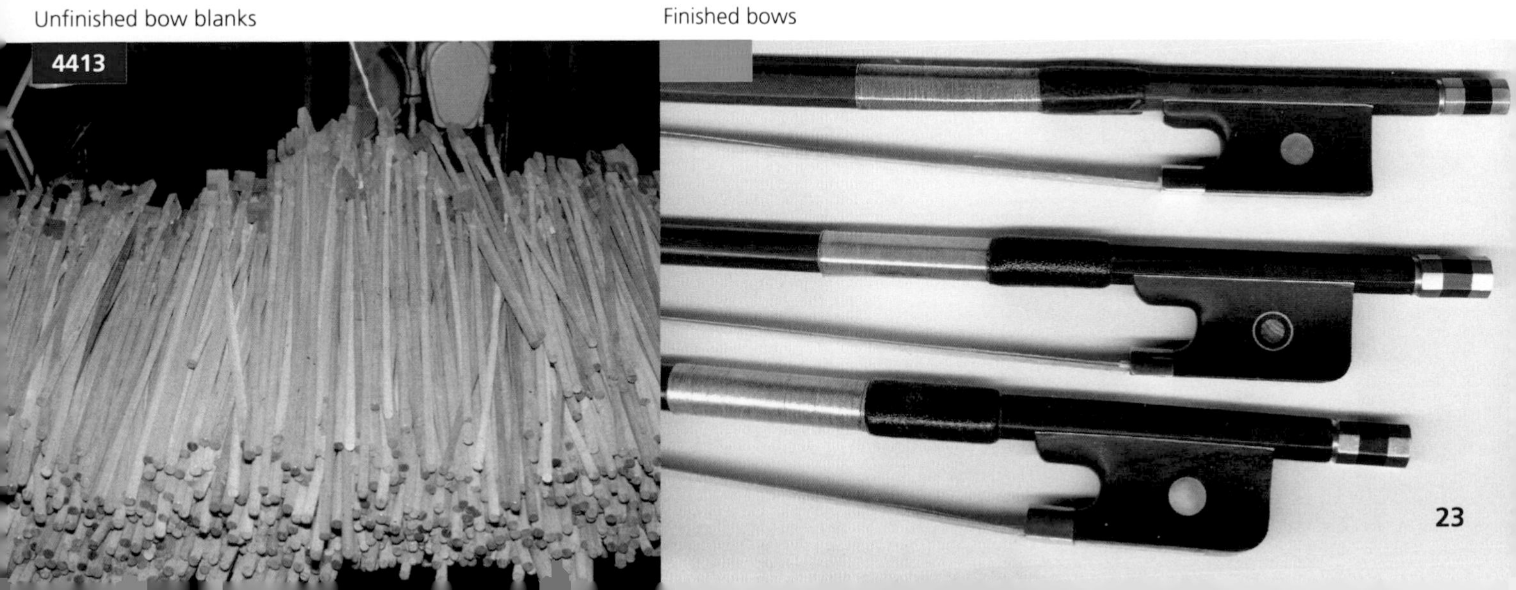

Cedrela
Spanish cedars

Distribution

The genus *Cedrela* consists of 17 species, three of which are listed in Appendix III / Annex C (*Cedrela fissilis, C. lilloi and C. odorata)* and four of which are regulated under Annex D only (*C. montana, C. oaxacensis, C. salvadorensis* and *C. tonduzii*). These species are native to Central and South America and the Caribbean including Argentina, Barbados, Belize, Bolivia, Brazil, Colombia, Costa Rica, Cuba, Dominica, Dominican Republic, Ecuador, El Salvador, French Guiana, Grenada, Guadeloupe, Guatemala, Guyana, Haiti, Honduras, Jamaica, Martinique, Mexico, Netherlands Antilles, Nicaragua, Panama, Paraguay, Peru, Puerto Rico, St Kitts and Nevis, St Lucia, Suriname, Trinidad and Tobago, Venezuela and the United Kingdom Overseas Territories (Cayman Islands, Monserrat) and the USA (Virgin Islands).

Uses

The species are traded for their workable, aromatic, insect-repelling and rot resistant timber which is used to make quality furniture, musical instruments, panelling, cigar boxes and in light construction (doors, boat building). Many old churches and cathedrals throughout the genus' distribution have panelling, doors and roofs made of *Cedrela*.

Trade

The CITES Trade Database indicates that the vast majority of trade relates to *Cedrela odorata* and is in sawn timber with lower levels of trade in timber pieces, carvings and veneer. The main exporters are Brazil, Bolivia, and Peru while Argentina, Mexico and the USA were the main importers followed by Canada, China, Japan, Spain, Puerto Rico and Paraguay. Imports into the EU are minimal for direct trade from exporting range states and there are discrepancies between trade levels reported by exporting and importing countries. The major EU importing member state is Spain followed by Denmark, France, Germany, the Netherlands, Sweden and the UK. There has been a small amount of trade in artificially propagated timber from plantations in Ghana and Cote d'Ivoire into the EU (Denmark and the UK). There is one recorded import of *C. fissilis* from Costa Rica to Spain, and no recorded trade for the four *Cedrela* species listed in Annex D.

Current Brazilian national legislation prohibits the domestic and international trade of all wild-collected native species (CITES and non-CITES species) with an IUCN Red List assessment of critically endangered (CR) and endangered (EN). All trade from Brazil should therefore only be in artificially propagated specimens.

The #5 annotation means only logs, sawn wood and veneer sheets are regulated. Definitions of these timber terms are found in the CITES Glossary (**http://www.cites.org/eng/resources/terms/glossary.php**) and in Resolution Conf. 10.13 (Rev. CoP15) (**http://www.cites.org/eng/res/10/10-13R15.php**). Finished products, such as musical instruments or furniture, are not regulated.

Plantations / artificial propagation

The species *Cedrela odorata* has been widely introduced outside its natural range as a shade tree for crops, an ornamental tree and for its timber. There are commercial plantations of this species producing timber for the international market in Africa (Ghana, Nigeria, Cote d'Ivoire, Madagascar, South Africa, Tanzania and Uganda) and in SE Asia (Viet Nam, the Philippines). Smaller or trial plantations have been established in Malaysia, Papua New Guinea and Thailand.

CITES international trade suspensions, export quotas and reservations

There are no current CITES international trade suspensions, export quotas or reservations in place for this genus.

There is a national export ban on commercial trade in Appendix II species from Paraguay (see Notification No. 2011/009 **http://www.cites.org/eng/notif/2011/E009.pdf**). This includes *C. odorata, C. lilloi* and *C. fissilis*.

EU decisions

There are no current EU suspensions or opinions for this genus.

See Species + for details http://www.speciesplus.net/#/taxon_concepts?taxonomy=cites_eu&taxon_concept_query=Cedrela&geo_entities_ids=&geo_entity_scope=cites&page=1

SCIENTIFIC AND COMMON NAMES	DATE OF LISTING	CURRENT LISTING AND ANNOTATION
Scientific names and authors: *Cedrela fissilis* Vell. *C. lilloi* C.DC. *C. odorata* L. *C. montana* Moritz ex Turcz. *C. oaxacensis* C.DC. & Rose *C. salvadorensis* Standl. *C. tonduzii* C.DC. **Family**: Meliaceae **Common names:** **English:** cedar, cigar-box cedar (*C. odorata*), red cedar, Spanish cedar, stinking mahogany **French:** acajou, cedrat, cedrela **German:** cedrela, westindsche zedar, westindsche scheinzedr, zigarenkistchenholz (*C. odorata*) **Italian:** cedro acjou (*C. odorata*) **Spanish:** cedrela, cedro, cedro blanco (*C. fissilis*), cedro colorado, cedro diamantina, cedro dulce (*C. tonduzii*), cedro pinta, cedro rosado (*C. odorata*), cedro rojo **CITES Standard Reference:** The CoP has adopted a standard reference for generic names (*The Plant Book*, 2nd Edition, D. J. Mabberley, 1997, Cambridge University Press reprinted with corrections 1998). Also see *A Monograph on the genus Cedrela* (T. D. Pennington & A. N. Muellner, 2010) for detailed information on scientific names and synonyms for this genus (http://www.kew.org/science/tropamerica/books.htm). If you require more formal guidance contact the CITES Secretariat.	**Appendix III** *C. fissilis*: 14/10/2010 (only the Bolivian population. All other populations in Annex D) *C. lilloi*: 14/02/2012 (only the Bolivian population. All other populations in Annex D) *C. odorata:* Brazil: 27/04/2011 Bolivia: 14/10/2010 In addition, the following countries listed their national populations: Colombia: 29/10/2001 Guatemala: 12/02/2008 Peru: 12/06/2001 **Annex C** Brazil – 14/02/2012 The Plurinational State of Bolivia: 14/02/2012 In addition, the following countries have listed their national populations: Colombia: 21/12/2001 Guatemala: 11/04/2008 Peru: 05/08/2001 **Annex D** *C. fissilis*: 11/04/2008 *C. lilloi*: 11/04/2008 *C. odorata*: 11/04/2008 (except for those populations listed in Annex C (Populations of Colombia, Guatemala and Peru). *C. montana*: 11/04/2008 *C. oaxacensis*: 11/04/2008 *C. salvadorensis*: 11/04/2008 *C. tonduzii*: 11/04/2008	**Appendix III** *C. fissilis*: 14/10/2010 *C. lilloi*: 14/10/2010 *C. odorata*: Brazil: 27/04/2011 Bolivia: 14/10/2010 In addition, the following countries listed their national populations: Colombia: 29/10/2001 Guatemala: 12/02/2008 Peru: 12/06/2001 **Annex C** *C. fissilis*: 20/12/2014 *C. lilloi*: 20/12/2014 *C. odorata*: Brazil: 20/12/2014 Bolivia: 20/12/2014 In addition, the following countries listed their national populations: Colombia: 20/12/2014 Guatemala: 20/12/2014 Peru: 20/12/2014 **Annex D** *C. montana*: 20/12/2014 *C. oaxacensis*: 20/12/2014 *C. salvadorensis*: 20/12/2014 *C. tonduzii*: 20/12/2014 **Annotation** **§4** (Annex D) Logs, sawn wood and veneer sheets **#5** Logs, sawn wood, and veneer sheets.

Cedrela fissilis Finished guitars Sawn wood

Dalbergia
Rosewoods

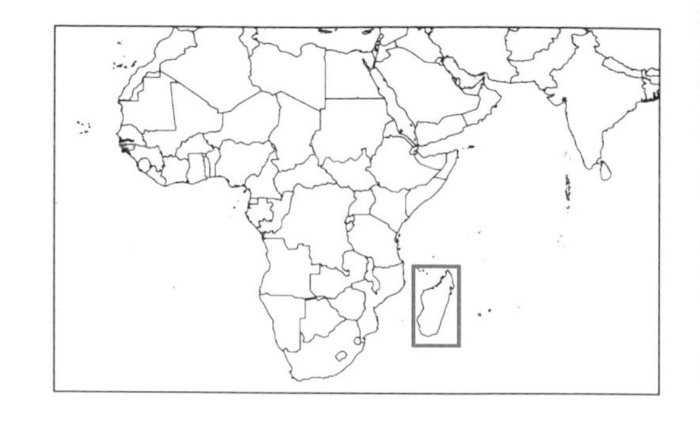

Distribution

The genus *Dalbergia* consists of approximately 250 species of trees, shrubs and vines which have a pan-tropical distribution. The CITES-listed *Dalbergia* species can be divided into three separate geographical areas – Africa (Madagascar), Central and South America, and Asia.

AFRICA (MADAGASCAR)

Distribution

There are currently 48 accepted species of Madgascan *Dalbergia* of which 47 are endemic (only found in Madagascar) and one, *D. bracteolata,* is also found in Kenya, Mozambique and Tanzania. The taxonomy of endemic Madagascan *Dalbergia* is under review so it is important to work from the current accepted list of scientific names (see Scientific Names below).

Uses

Dalbergia species are traditionally used locally for house construction, fences and woodworking. However, timber from Madagascan *Dalbergia* species are now among the species classified under China's National Hongmu Standard for use in the manufacture of luxury deep red coloured Hongmu furniture, highly sought after in Asia, in particular China. The timber from these species is also used in the manufacture of musical instruments, in particular guitars. China is the only country that has a specific customs code for Hongmu species – 44039930.

Trade

The recent listing of these species in 2013 means there is very little trade data available. All trade is for scientific purposes from Madagascar to Switzerland except for one pre-Convention commercial export of 36,000kg of *Dalbergia louvelii* from Madagascar to China in 2012. There are also three pre-Convention exports of very small amounts of *D.baronii* and *Dalbergia* spp. to Japan and Ecuador in 2013. However, demand is high as evidenced by large seizures of illegally logged and exported *Dalbergia* timber from Madagascar en route to China.

The #5 annotation means only logs, sawn wood and veneer sheets are regulated. Definitions of these product terms are found in the CITES Glossary (**http://www.cites.org/eng/resources/terms/glossary.php**) and in Resolution Conf. 10.13 (Rev. CoP15) (**http://www.cites.org/eng/res/10/10-13R15.php**). Finished products, such as musical instruments or furniture, are not regulated. The listing of the Madagascan populations of *Dalbergia* species means that only regulated products of a *Dalbergia* species endemic to Madagascar are covered under this listing.

Guitars

Logs buried on a beach

Plantations / artificial propagation

There are no known commercial plantations or artificial propagation of these species in Madagascar.

CITES international trade suspensions, export quotas and reservations

There are no current CITES international trade suspensions in place for these species.

There is a current zero export quota in place for the *Dalbergia* populations of Madagascar.

There is a current reservation in place for this listing.

EU Decisions

There are no current EU suspensions or opinions for these species.

See Species + for details **http://www.speciesplus.net/#/taxon_concepts?taxonomy=cites_eu&taxon_concept_query=Dalbergia&geo_entities_ids=&geo_entity_scope=cites&page=1**

SCIENTIFIC AND COMMON NAMES	DATE OF LISTING	CURRENT LISTING AND ANNOTATION
Scientific name and author: *Dalbergia* spp. (populations of Madagascar only). See **http://www.cites.org/sites/default/files/eng/com/sc/65/Inf/E-SC65-Inf-21.pdf** for individual species names and authors	**Appendix III** Five Malagasy species only (*D. louvelii, D. monticola, D. normandii, D. purpurascens* and *D. xerophila*): 22/12/2011	
Family: Leguminosae	**Annex C** *D. louvelii, D. monteola, D. normandii, D. purpurascens, D. xerophila:* 15/12/2012	
Common names: **English:** rosewoods, palisanders **French:** bois de rose **German:** palisander **Italian:** palissandro; legno di rose **Malagasy:** andramena, bolabola, hazoambo, hazovola mena, hendramena, hitsika, manary, manary mainty, manjakabenitany, sovoka, tombobitsy, tongobitsy, voamboana, volombodipona	**Appendix II** *Dalbergia* spp. (populations of Madagascar only): 12/06/2013 **Annex B** *Dalbergia* spp. (populations of Madagascar only): 10/08/2013	**Appendix II** *Dalbergia* spp. (populations of Madagascar only): 12/06/2013 **Annex B** *Dalbergia* spp. (populations of Madagascar only): 20/12/2014
CITES Standard Reference: The CoP has adopted a standard reference for generic names (*The Plant-Book*, second edition, D. J. Mabberley, 1997, Cambridge University Press reprinted with corrections 1998). A preliminary checklist for Madagascan *Dalbergia* species for use by the CITES Parties was adopted at the 65th Standing Committee (July 2014). See **http://www.cites.org/sites/default/files/eng/com/sc/65/Inf/E-SC65-Inf-21.pdf.** To check if updates have been made to this list or if you require more formal guidance contact the CITES Secretariat.		**Annotation** #5 Logs, sawn wood and veneer sheets (populations of Madagascar only)

Sri Lankan seizure of logs

Rosewood log

Dalbergia Rosewoods

(continued)

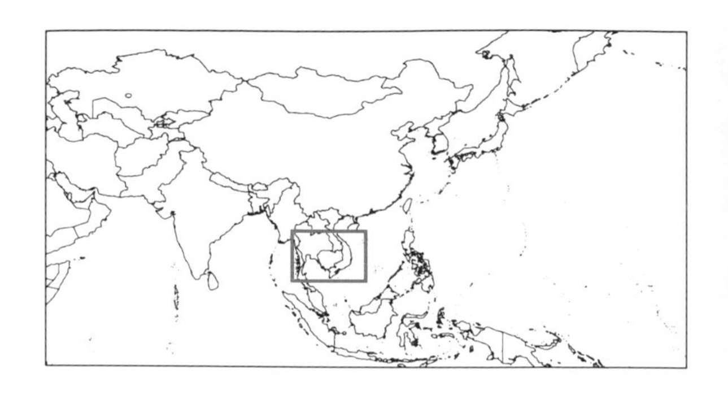

ASIA

Distribution

The only Asian species of *Dalbergia* currently listed under CITES is *Dalbergia cochinchinensis*. It is native to Cambodia, Lao People's Democratic Republic, Thailand and Viet Nam.

Uses

The timber has a red colouration highly sought after in Asia, in particular China. This species is among the timber species classified under China's National Hongmu Standard for use in the manufacture of luxury Hongmu furniture. China is the only country that has a specific customs code for Hongmu species – 44039930. The timber is also used to make musical instruments (lutes and Erhu Chinese violins which are sometimes made from old furniture), jewellery, gun blanks, sewing-machines, sports equipment and as a decorative veneer (e.g. in passenger ships and for instrument cases). However, given the current high prices fetched by this species its trade is for high end products rather than for local consumption or low end production.

Trade

The recent listing of this species in 2013 means there is no trade data available. However, the trade in timber (logs, sawn wood and some veneers) is evidenced by seizures and NGO campaigns on the illegal trade in this species. Destination markets are mainland China, Hong Kong SAR, Singapore and Taiwan. Prices for timber vary from \$6,000-50,000/m^3. Timber may be traded under the name *Dalbergia cambodiana* but this is a synonym of *Dalbergia cochinchinensis*. As such, specimens traded under the name *Dalbergia cambodiana* are included in Appendix II, and such trade should be regulated accordingly (**http://cites.org/sites/default/files/notif/E-Notif-2014-061.pdf**) The common name used for this species (rosewood) may refer to other non-CITES listed species of Asian *Dalbergia* (*D. bariensis*) or species from other genera (*Pterocarpus macrocarpus*, synonym *P. pedatus*).

The #5 annotation means only logs, sawn wood and veneer sheets are currently regulated. Definitions of these timber terms are found in the CITES Glossary (**http://www.cites.org/eng/resources/terms/glossary.php**) and in Resolution Conf. 10.13 (Rev. CoP15) (**http://www.cites.org/eng/res/10/10-13R15.php**). Finished products, such as musical instruments or furniture, are not regulated.

Hongmu tables

Bed

Chair carving

Plantations / artificial propagation

No commercial plantations of this species exist. Therefore all trade will be wild in origin. As this species is slow growing the artificial propagation trials that have been encouraged in range states will take many years to produce sizeable timber for export.

CITES international trade suspensions, export quotas and reservations

There is a current CITES international trade suspension for this species from Lao People's Democratic Republic. There are no export quotas or reservations in place for this species. All range states have domestic legislation in place to ban the logging and/or export of this species.

EU Decisions

There are no current EU opinions for this species

See Species + for details **http://www.speciesplus.net/#/taxon_concepts/15804/legal**

SCIENTIFIC AND COMMON NAMES	DATE OF LISTING	CURRENT LISTING AND ANNOTATION
Scientific name and author: *D. cochinchinensis* Pierre (may be cited as *D. cochinchinensis* Laness.)	**Appendix II** 12/06/2013	**Appendix II** 12/06/2013
Family: Leguminosae	**Annex B** 10/08/2013	**Annex B** 20/12/2014
Common names: **Cambodia:** kra nhoung, kranhung cheam moan (locally known as black-red-stripe wood) **Chinese:** hongmu, hongsuanzhi 紅酸枝 **English:** rosewood, Siamese rosewood, Thailand rosewood, Vietnamese rosewood, redwood trắc wood **Lao:** mai khanhoung, mai khayung **Thai:** mai payoong **Vietnamese:** cẩm lai, trac (tracwood)		**Annotation** #5 Logs, sawn wood, and veneer sheets.
CITES Standard Reference: The CoP has adopted a standard reference for generic names (*The Plant-Book*, second edition, D. J. Mabberley, 1997, Cambridge University Press reprinted with corrections 1998). The Plant List is a useful source of information on scientific names and synonyms of this species (**http://www.theplantlist.org/tpl1.1/record/ild-46412**). If you require more formal guidance contact the CITES Secretariat.		

Logs

Logs

Unloading Hongmu furniture

Dalbergia
Rosewoods

(continued)

CENTRAL AND SOUTH AMERICA

Distribution

The Central and South American CITES-listed *Dalbergia* species are *D. calycina* (Belize, Costa Rica, El Salvador, Guatemala, Mexico, Nicaragua); *D. cubilquitzensis* (Costa Rica, Mexico, Nicaragua); *D. darienensis* (Colombia and Panama); *D. glomerata* (Costa Rica, Guatemala, Honduras, Mexico); *D. granadillo* (El Salvador and Mexico); *D. nigra* (Brazil); *D. retusa* (Belize, Costa Rica, El Salvador, Guatemala, Honduras, Mexico, Nicaragua and Panama); *D. stevensonii* (southern Belize and nearby regions of Guatemala and Mexico); and *D. tucurensis* (Guatemala and Nicaragua). For some of these species only certain populations are regulated and for others their listing under the EU Wildlife Trade Regulations has not yet come into force (see Species + for individual distribution and listing details **http://www.speciesplus.net/species#/taxon_concepts?taxonomy=cites_eu&taxon_concept_query=Dalbergia&geo_entities_ids=&geo_entity_scope=cites&page=1**).

Uses

Uses vary from species to species but include: *Dalbergia cubilquitzensis* (chess sets, general carpentry, jewellery boxes, luxury furniture, musical instruments, tool handles, veneer, handicrafts); *D. darienensis* (cabinet making, furniture, marquetry, parquet flooring, musical instruments); *D. glomerata* (furniture and general construction); *D. nigra* (furniture, art work, musical instruments particularly guitars); *D. granadillo / D. retusa* (inlay work, scientific instruments, tool and cutlery handles, hand gun grips, butts of billiard cues, decorative and figured veneers, parquet floors, car dashboards, jewellery boxes, canes and in particular guitars and woodwind instruments such as clarinets); *D. stevensonii* (percussion bars for marimbas and xylophones, fingerboards for banjos, guitars and mandolins, harp bodies, furniture and decorative veneer, particularly in the Asian market); and *D. tucurensis* (guitars, boxes, humidors, picture frames, marimbas, guitars, ukuleles).

Trade

Given the majority of these species are newly listed in Appendix II / Annex B or Appendix III / Annex C there is little trade data available for all but a few species i.e. *D. stevensonii, D. retusa* and *D. nigra*. Trade levels are often difficult to interpret given the many different terms and units cited for some species and products (e.g. kg, m^3, carvings) or there is little differentiation between the species as trade is recorded as "*Dalbergia* spp." or "rosewoods".

Dalbergia nigra table | *D. stevensonii* logs | Chess set

The main exporters of timber of *Dalbergia stevensonii* are Guatemala and Belize with Germany, China and the USA as the major importers. This species is re-exported from the EU (Germany) to Japan, USA and Turkey. For *D. retusa / D. granadillo* (included as may be traded under both species names and supposedly indistinguishable from *D. retusa*) there is little trade data available but the trade appears to be in timber from Guatemala, Mexico, Nicaragua and Panama. The major importers are Spain, the USA, Switzerland and China (including Hong Kong). Re-exports are of timber and small high value items such as gun blanks.

For *Dalbergia nigra* the species is listed in Appendix I / Annex A and the international trade in wild-sourced specimens for commercial purposes is prohibited. Commercial trade in artificially propagated specimens is permitted. The CITES Trade Database shows that the majority of trade has been in specimens harvested prior to the species listing in 1992 i.e. pre-Convention material. However, the EU does not fully implement the exemptions for commercial use of Appendix I listed species (as laid out in the Convention text) through its Wildlife Trade Regulations (Council Regulation 338/97 – see *KEY RESOURCES* page for reports and websites containing more information). The trade in this species is in a variety of units including carvings, sawn wood logs, timber, timber pieces and veneer and trade in wild-sourced specimens has taken place. The trade is also shifting to processed products (furniture, carvings in particular musical instruments and veneer) and from direct exports (from the only range state, Brazil) to re-exports of pre-Convention material. The largest importers of this material are Japan, the USA, Canada, Malaysia and the EU (Spain, the Netherlands, the UK, Italy and Germany). The main re-exporters are the USA, the EU (Spain, Germany and the UK). Current Brazilian national legislation prohibits the domestic and international trade of all wild-collected native species (CITES and non-CITES species) with an IUCN Red List assessment of critically endangered (CR) and endangered (EN). All trade from Brazil should therefore only be in artificially propagated specimens.

All but one of the Central and South American *Dalbergia* species have a #6 annotation. This means that only logs, sawn wood, veneer sheets and plywood are regulated. Definitions of these timber terms are found in the CITES Glossary (**http://www.cites.org/eng/resources/terms/glossary.php**) and in Resolution Conf. 10.13 (Rev. CoP15) (**http://www.cites.org/eng/ res/10/10-13R15.php**). Under this annotation finished products, such as musical instruments or furniture, are not regulated. The annotation for *D. darienensis* is #2, which means all parts and derivatives except seeds, pollen and finished products packaged and ready for retail trade are regulated. As Appendix I / Annex A listed (*D. nigra*) all parts and derivatives, live or dead, are regulated.

Plantations / artificial propagation

Tree planting schemes are in place for some species but the majority of trade will be wild in origin. Planting schemes for *D. stevensonii* have been established in Belize (following damage caused by Hurricane Iris in 2001) and *D.retusa/granadillo* plantations are maintained in Costa Rica and Nicaragua although the majority of wood still comes from privately owned "fincas" (rural or agricultural land that was planted 80–100 years ago). There are no known plantations or artificial propagation of *D. nigra*.

CITES international trade suspensions, export quotas and reservations

There are no current CITES international trade suspensions and export quotas for these species.

Seized *Dalbergia retusa* sawn wood — Gun blank

EU Decisions

There are no current EU suspensions or opinions for these species.

See Species + for details http://www.speciesplus.net/species#/taxon_concepts?taxonomy=cites_eu&taxon_concept_query=Dalbergia&geo_entities_ids=&geo_entity_scope=cites&page=1

SCIENTIFIC AND COMMON NAMES	DATE OF LISTING	CURRENT LISTING AND ANNOTATION
Scientific names and authors: *Dalbergia calycina* Benth. *Dalbergia cubilquitzensis* (Donn.Sm.) Pittier *Dalbergia darienensis* Rudd *Dalbergia glomerata* Hemsl. *Dalbergia granadillo* Pittier *Dalbergia nigra* (Vell.) Benth. *Dalbergia retusa* Hemsl. *Dalbergia stevensonii* Standl. *Dalbergia tucurensis* Donn.Sm. **Family**: Leguminosae **Common names** **See Species+ for individual species common names** English: black rosewood, rosewood, Nicaraguan rosewood, Guatemalan rosewood (*D. cubilquitzensis)* Honduran rosewood; Yucatan rosewood (*D. tucurensis*), Panama rosewood French: palissandre du Honduras, palissandre cocobolo German: coboboloholz, foseholz Portugese: pau-preto (only for *D. nigra*), jacarandá, cocobolo (*D. retusa*) Spanish: cocobolo, cocobolo negro, granadillo, granadillo de Chontales, franadillo, funera, namba, ñambar, palissandre du Honduras, palisandro, palisandro de Honduras **CITES Standard Reference:** The CoP has adopted a standard reference for generic names (*The Plant Book*, second edition, D. J. Mabberley, 1997, Cambridge University Press reprinted with corrections 1998). The Plant List is a useful source of information on scientific names and synonyms of these species (http://www.theplantlist.org/tpl1.1/search?q=Dalbergia). If you require more formal guidance contact the CITES Secretariat.	**Appendix I** *D. nigra*: 11/06/1992 **Annex A** *D. nigra*: 01/06/1997 **Appendix III** *D. calycina*: 05/02/2015 (population of Guatemala) *D. cubilquitzensis*: 05/02/2015 (population of Guatemala) *D. darienensis*: 22/12/2011 (population of Panama) *D. glomerata*: 05/02/2015 (population of Guatemala) *D. retusa*: 12/02/2008 (population of Guatemala) *D. stevensonii*: 12/03/2008 (population of Guatemala) *D. tucurensis*: 24/06/2014 **Annex C** *D. darienensis*: 15/12/2012 *D. tucurensis*: 20/12/2014 *D. calycina, D. cubilquitzensis* and *D. glomerata* currently not listed in EU Annexes **Appendix II** *D. granadillo*: 12/06/2013 **Annex D** *D. granadillo*: 11/04/2008 *D. retusa*: 11/04/2008 (all populations except those of Guatemala on Annex C) *D. stevensonii*: 11/04/2008 (all populations except those of Guatemala on Annex C)	**Appendix I** *D. nigra*: 11/06/1992 **Annex A** *D. nigra*: 20/12/2014 **Appendix III** *D. calycina*: 05/02/2015 (population of Guatemala) *D. cubilquitzensis*: 05/02/2015 (population of Guatemala) *D. darienensis*: 12/06/2013 (population of Panama) *D. glomerata*: 05/02/2015 (population of Guatemala) *D. tucurensis*: 05/02/2015 (population of Guatemala) **Annex C** *D. darienensis*: 20/12/2014 *D. tucurensis*: 20/12/2014 **Appendix II** *D. granadillo*: 12/06/2013 *D. retusa*: 12/06/2013 *D. stevensonii*: 12/06/2013 **Annex B** *D. granadillo*: 20/12/2014 *D. retusa*: 20/12/2014 *D. stevensonii*: 20/12/2014 **Annotation** As **Appendix I / Annex A** listed (*D. nigra*) all parts and derivatives, live or dead, are regulated. **#6** Logs, sawn wood, veneer sheets and plywood. **#2** All parts and derivatives, except: a) seeds and pollen; and b) finished products packaged and ready for retail trade

Safmarin
MAERS
MAERS
CXDU 218052 0
45G1
TCLU 159689 1
45G1
TAL
CXDU 214999 3
45G1
tex
TGHU 627551 1
45G1
MAX. WT.
TARE-WT.
PAYLOAD
CU. CAP.
CARU 965609 2
45G1
msc
MEDU 402691 6
42G1
M.G.W.
TARE
NET
CU.CAP.
msc
MSCU 479867 8
42G1
M.G.W.
TARE
NET
CU.CAP.
FLORENS
www.florens.com
FSCU 434763 5
42G1
MAX. GROSS
MAX. CARGO
CU. CAP.
msc
MSCU 494482 8
42G1
M.G.W.
TARE
NET
CU.CAP.
BEACON

Diospyros
Ebonies

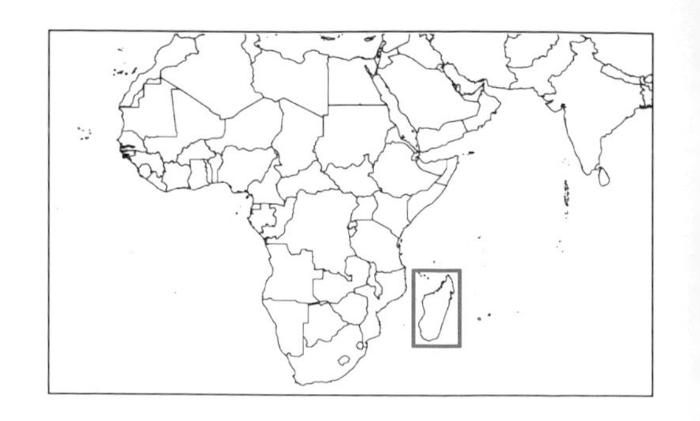

Distribution

Over 500 species have been described in the genus *Diospyros* but many of the species names in circulation are incorrect because they may, amongst other things, represent more than one species. The taxonomy of the genus is currently in revision. The number of ebony species in Madagascar ranges from 85 to 240, but currently only 87 species are recognised as valid names. All but one of these species, *D. ferrea* which is also found along the coasts of East Africa and India, are endemic to Madagascar. The listing of the Madagascan populations of *Diospyros* species means that only regulated products of a *Diospyros* species endemic to Madagascar is regulated.

Uses

Ebony is used in the manufacture of furniture, musical instruments, in particular fingerboards for violins and guitars, piano keys, violin/viola pegs, wind instruments, cutlery and marquetry and inlay. Highly valuable ebony species include *Diospyros gracilipes*, *D. perrieri*, and *D. platycalyx*.

Trade

The recent listing of these species in Appendix II in 2013 means there is very little trade data available. According to the CITES Trade Database, there is only a small amount of trade for scientific purposes to Switzerland, Germany and the UK. It appears that *D. perrieri* and *D. tropofila* are in trade.

The #5 annotation means only logs, sawn wood and veneer sheets are regulated. Definitions of these timber terms are found in the CITES Glossary (**http://www.cites.org/eng/resources/terms/glossary.php**) and in Resolution Conf. 10.13 (Rev. CoP15) (**http://www.cites.org/eng/res/10/10-13R15.php**). Finished products, such as musical instruments or furniture, are not regulated. The listing of *Diospyros* species (populations of Madagascar) means that only regulated specimens of *Diospyros* species endemic to Madagascar are regulated.

Plantations / artificial propagation

There are no known plantations or artificial propagation of these species within Madagascar.

Diospyros sp. Ebony ornaments Ebony inlay

CITES international trade suspensions, export quotas and reservations

There are no current CITES international trade suspensions in place for *Diospyros* species.

There is a current zero export quota for *Diospyros* material exported from Madagascan populations only.

There are current reservations in place for these species.

EU Decisions

There are no current EU suspensions or opinions for Madagascan *Diospyros* species.

See Species + for details
http://www.speciesplus.net/#/taxon_concepts?taxonomy=cites_eu&taxon_concept_query=Diospyros&geo_entities_ids=&geo_entity_scope=cites&page=1

SCIENTIFIC AND COMMON NAMES	DATE OF LISTING	CURRENT LISTING AND ANNOTATION
Scientific names and authors: *Diospyros* L. (populations of Madagascar only). See Species + and CITES Standard Reference below for individual names and authors	**Appendix III** 104 species of *Diospyros* (populations of Madagascar only) listed: 22/12/2011	**Appendix II** Genus listing *Diospyros* spp. (populations of Madagascar only): 12/06/2013
Family: Ebenaceae		
Common names: **English:** ebony, ebonies **French:** bois d'ébène **German:** ebenholz **Italian:** ebano **Malagasy:** kakazomainty, hazomafana, hazomainty, lopingo, maintipody, maintipototra, mapingo, pingo **Portugese:** ébano	**Appendix II** Genus listing *Diospyros* spp. (populations of Madagascar only): 12/06/2013	**Annex B** Genus listing *Diospyros* spp. (populations of Madagascar only): 20/12/2014
CITES Standard Reference: The CoP has adopted a standard reference for generic names (*The Plant Book*, second edition, D. J. Mabberley, 1997, Cambridge University Press reprinted with corrections 1998). A list of Madagascan *Diospyros* species can be downloaded from the Catalogue of the Vascular Plants of Madagascar (**http://www.tropicos.org/NameSearch.aspx?projectid=17&name=Diospyros**). If you require more formal guidance contact the CITES Secretariat.	**Annex C** 104 species of *Diospyros* (populations of Madagascar only) listed: 15/12/2012	**Annotation** #5 Logs, sawn wood and veneer sheets (populations of Madagascar only)

Violin pegs

Violin with ebony fingerboard

Piano with ebony keys

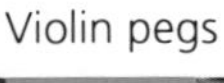

Dipteryx panamensis
Almendro

Distribution

This large tree can grow to 60 metres in height and is found in the humid rainforests of southern Nicaragua, Costa Rica, Panama, and Colombia.

Uses

This species has one of the heaviest woods in the world, and due to its mechanical resistance it was difficult to harvest before the advances in chainsaw technology in the 1980s. It is used in marine construction, sports equipment and construction of railroad ties, bridges, industrial flooring, boats, springboards, industrial machinery, and agricultural tool handles.

Trade

The CITES Trade Database shows little trade with limited exports of timber, the majority of which is exported by Panama then smaller volumes by Costa Rica and Nicaragua. The trade is in wild specimens and the main importer is the USA. The main EU importer is Germany. In 2008 the harvesting of almendro trees from the wild was completely banned in Costa Rica. The taxonomy of this listing is in need of revision given the species name *Dipteryx panamensis* is commonly believed to be a synonym of *D. oleifera*.

There is no annotation therefore all parts and derivatives, live or dead, are regulated.

Plantations / artificial propagation

There are some commercial and experimental plantations of this species in Costa Rica, but the production levels of these are unknown.

Boat – marine construction

Tool handles

CITES international trade suspensions, export quotas and reservations

There are no current CITES international trade suspensions, export quotas or reservations in place for this species.

EU Decisions

There are no current EU suspensions or opinions for this species.

See Species + for details **http://www.speciesplus.net/#/taxon_concepts/22966/legal**

SCIENTIFIC AND COMMON NAMES	DATE OF LISTING	CURRENT LISTING AND ANNOTATION
Scientific name and author: *Dipteryx panamensis* Pittier Record & Mell *Dipteryx oleifera* Benth.	**Appendix III** Costa Rica: 13/02/2003 Nicaragua: 13/09/2007	**Appendix III** Costa Rica: 13/02/2003 Nicaragua:13/09/2007
Family: Leguminosae		
Common names: English: almendro, tonka bean tree German: waldmandelbaum Spanish: almendro	**Annex C** Costa Rica: 30/08/2003 Nicaragua: 11/04/2008	**Annex C** Costa Rica: 20/12/2014 Nicaragua: 20/12/2014 **Annotation** No annotation is given therefore all parts and derivatives, live or dead, are regulated.
CITES Standard Reference: The CoP has adopted a standard reference for generic names (*The Plant Book*, 2nd Edition, D. J. Mabberley, 1997, Cambridge University Press reprinted with corrections 1998). The Plant List is a useful source of information on scientific names and synonyms for this species **http://www.theplantlist.org/tpl1.1/record/ild-40221** and **http://www.theplantlist.org/tpl1.1/record/ild-10283** If you require more formal guidance contact the CITES Secretariat.		

Sports equipment

Fitzroya cupressoides
Alerce

Distribution

This species of conifer is one of the largest South American trees, reaching over 50 metres in height, and it is the only species in the genus *Fitzroya*. It is native to southern Argentina and southern Chile and can live to over 2,000 years.

Uses

The reddish brown, straight grained, insect resistant timber is durable and easily worked, and has been used in the manufacture of musical instruments, furniture, roof shingles and ships masts.

Trade

This species is listed in Appendix I / Annex A and the international trade in wild-sourced specimens for commercial purposes is prohibited. Commercial trade in artificially propagated specimens is permitted. The CITES Trade Database shows that the majority of trade is in pre-Convention timber worked into carvings with the main exporter being Chile. The main importers of these products are the USA and Europe (France and Spain). Between 2003–2004 around 70m^3 of wild-collected timber was exported from Chile to Taiwan. This trade may have been misreported pre-Convention timber. Artificially propagated live plants and seeds are for sale on the Internet and through plant nurseries.

As an Appendix 1/Annex A listed species there is no annotation therefore all parts and derivatives, live or dead, are regulated.

Plantations / artificial propagation

There are no known commercial plantations of this species and logging is illegal but the species was introduced into cultivation in Europe in 1849 so large specimens may be found in parks and gardens.

Sawn wood

Boxes

CITES international trade suspensions, export quotas and reservations

There are no current CITES international trade suspensions, export quotas or reservations in place for this species.

EU Decisions

There are no current EU suspensions or opinions for this species.

See Species + for details **http://speciesplus.net/#/taxon_concepts/27947/legal**

SCIENTIFIC AND COMMON NAMES	DATE OF LISTING	CURRENT LISTING AND ANNOTATION
Scientific name and author: *Fitzroya cupressoides* (Molina) I.M.Johnston	**Appendix I** 01/07/1975 **Annex A** 01/06/1997	**Appendix I** 22/10/1987 (all populations) **Annex A** 20/12/2014 **Annotation** For Appendix I/Annex A species all parts and derivatives, live or dead, are regulated
Family: Cupressaceae		
Common names: **English:** Patagonian cypress, Chilean false larch **French:** bois d'Alerce, cyprès de Patagonie, fitzroia **Italian:** cipresso della Patagonia **Portugese:** ciprés da Patagónia **Spanish:** ciprés de la Patagonia, alerce, falso alerce Chileno, lahual, lahuan		
CITES Standard Reference: The CoP has adopted a standard reference for generic names (*The Plant Book*, 2nd Edition, D. J. Mabberley, 1997, Cambridge University Press reprinted with corrections 1998). The Plant List is a useful source of information on scientific names and synonyms for this species (**http://www.theplantlist.org/tpl1.1/record/kew-2813201**). If you require more formal guidance contact the CITES Secretariat.		

House shingles

Sawn wood

Fraxinus mandshurica
Manchurian ash

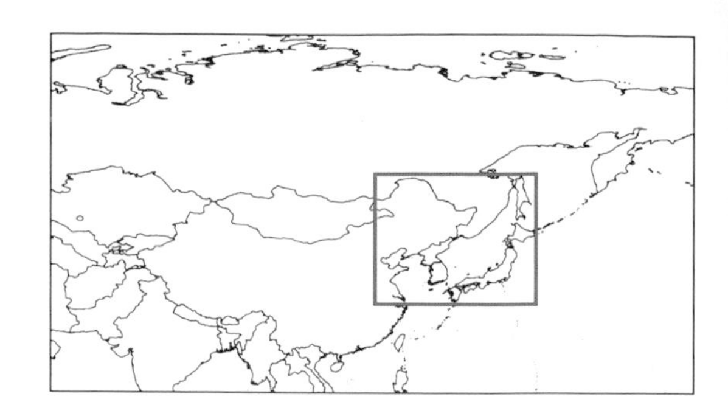

Distribution

This is currently the only species in the genus *Fraxinus* (approximately 58 species) listed under CITES. The species is native to northern China, Republic of Korea, Japan and south east Russian Federation (Sakhalin Island).

Uses

An important and valuable hardwood tree, its timber is used in the manufacture of furniture, utensils, sporting goods and veneers.

Trade

The recent listing of this species in 2014 means there is no trade data available. Given plantations exist both wild and plantation grown material may be in trade.

The #5 annotation means only logs, sawn wood and veneer sheets are regulated. Definitions of these product terms are found in the CITES Glossary (**http://www.cites.org/eng/resources/terms/glossary.php**) and in Resolution Conf. 10.13 (Rev. CoP15) (**http://www.cites.org/eng/res/10/10-13R15.php**). Finished products, such as furniture, are not regulated.

Plantations / artificial propagation

Commercial plantations of this species exist in China and this species has been planted as an ornamental tree in Canada, the USA and Europe.

Furniture

Utensils

Veneer sheet

CITES international trade suspensions, export quotas and reservations

There are no current CITES international trade suspensions, export quotas or reservations in place for this species.

EU Decisions

There are no current EU suspensions or opinions for this species.

See Species + for details **http://speciesplus.net/#/taxon_concepts/65604/legal**

SCIENTIFIC AND COMMON NAMES	DATE OF LISTING	CURRENT LISTING AND ANNOTATION
Scientific name and author: *Fraxinus mandshurica* Rupr.	**Appendix III** 24/06/2014 **Annex C** 20/12/2014	**Appendix III** 24/06/2014 **Annex C** 20/12/2014 **Annotation** #5 Logs, sawn wood and veneer sheets
Family: Oleaceae		
Common names: **English:** Manchurian ash **German:** Mandschurische esche **Japanese:** Yachidamo **Portugese:** freixo da Manchúria **Swedish:** manchurisk ask		
CITES Standard Reference: The CoP has adopted a standard reference for generic names (*The Plant Book*, 2nd Edition, D. J. Mabberley, 1997, Cambridge University Press reprinted with corrections 1998). The Plant List is a useful source of information on scientific names and synonyms for this species (**http://www.theplantlist.org/tpl1.1/record/kew-369861**). If you require more formal guidance contact the CITES Secretariat.		

Gonystylus
Ramin

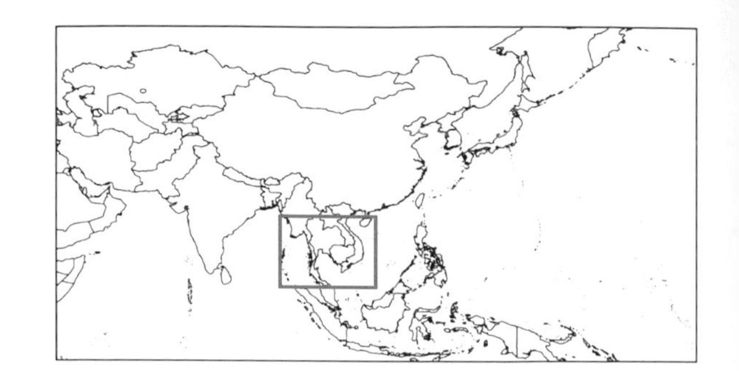

Distribution

Ramin is the common name given to all species in the genus *Gonystylus* which is listed under CITES. The 30 or so species are native to the peat swamp forests of southeast Asia, including Brunei Darussalam, Fiji, Indonesia, (Kalimantan and Sumatra) Malaysia (Peninsular Malaysia, Sabah and Sarawak), Singapore, the Solomon Islands and the Philippines.

Uses

The light coloured, easily turned wood is in trade as a wide range of semi-finished and finished products. These include mouldings, dowels, handles, paint brush blanks, finished paint brushes, curtain rods, umbrella poles, sports equipment (snooker and pool cues), toys (dolls house miniatures), furniture, cots, tool handles, technical drawing implements, window shutters, slatted wooden blinds, picture frames, slatted louvre wooden doors and veneers.

Trade

The key ramin products in trade are sawn timber, timber pieces and carvings. All trade is wild in origin and trade is reported at the genus or species level with *Gonystylus bancanus* being the dominant species in trade. According to the CITES Trade Database, Malaysia is the main exporter with smaller volumes coming from Indonesia. Lower levels of exports are also reported from non-range states (China, India, Singapore, Taiwan and Viet Nam). Major importers include Malaysia, the EU (dominated by the Netherlands and Italy), China, Japan, Switzerland and the USA. The trade data also shows that re-exports of ramin, in particular sawn wood, carvings and timber pieces, were reported by China, the EU (Germany, Italy, Spain and the UK), India, Japan and Singapore. There are discrepancies between exporter and importer data and trade in a product may be reported in different units (kg or m^3).

The #4 annotation means all parts and derivatives are regulated apart from seeds, pollen, tissue cultured plants in sterile containers and cut flowers from artificially propagated plants. Therefore all finished and semi-finished products, such as picture frames, paint brushes, blinds and snooker cues, are regulated.

Plantations / artificial propagation

There are no known commercial plantations or artificial propagation of this genus so all trade is wild in origin.

Logs

Sawn wood

CITES international trade suspensions, export quotas and reservations

There are no current CITES international trade suspensions in place for this genus.

There is a current reservation for this genus.

There are current export quotas in place for ramin species from Indonesia and Malaysia. Check Species + for details or check the annual export quotas on the CITES website (http://www.cites.org/eng/resources/quotas/index.php).

EU Decisions

There are current EU opinions in place for certain ramin species/country combinations

See Species + for details http://www.speciesplus.net/#/taxon_concepts?taxonomy=cites_eu&taxon_concept_query=Gonystylus&geo_entities_ids=&geo_entity_scope=cites&page=1

Mouldings

Mouldings – picture frames

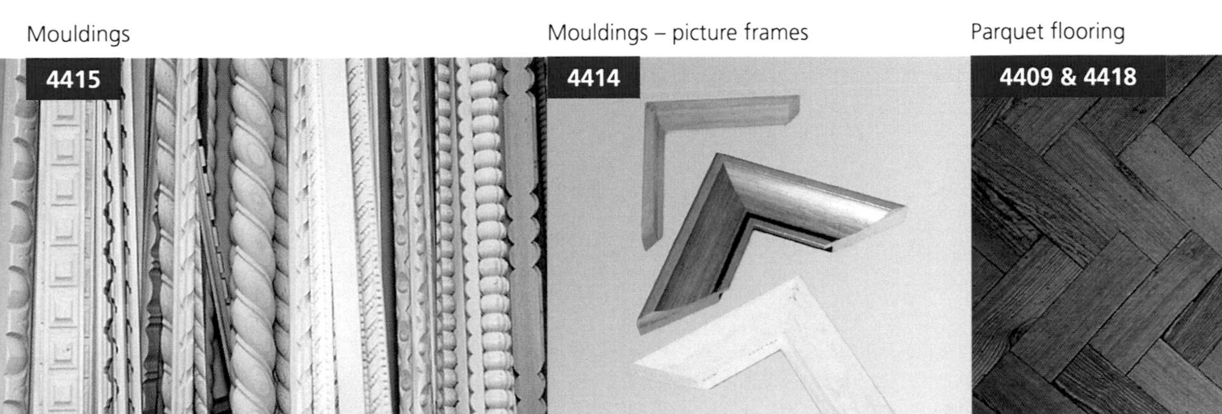

Parquet flooring

Mouldings

Veneer sheets

Picture frames

Garden tool handles

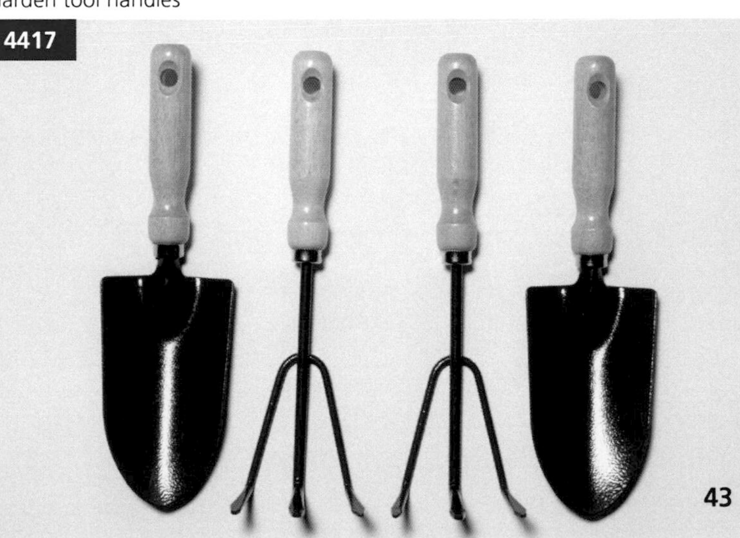

SCIENTIFIC AND COMMON NAMES	DATE OF LISTING	CURRENT LISTING AND ANNOTATION
Scientific name and author: *Gonystylus* See Species + for individual species names and authors **Family**: Thymelaeaceae **Common names:** **English:** ramin **French:** ramin **Indonesian:** rahara buaya (Sumatra, Kalimantan), medang keladi (Kalminatan) **Italian:** ramino **Malaysian:** gahara buaya (Sarawak), melawis, ahmin, kaya garu, ramin telur **Spanish:** ramin **Swedish:** raminsläktet **CITES Standard Reference:** The CoP has adopted a standard reference for generic names (*The Plant Book*, Second Edition, D. J. Mabberley, 1997, Cambridge University Press reprinted with corrections 1998). The Plant List is a useful source of information on scientific names and synonyms of the species in this genus (**http://www.theplantlist.org/tpl1.1/search?q=gonystylus**). If you require more formal guidance contact the CITES Secretariat.	**Appendix III** 06/08/2001 (genus listing for *Gonystylus* spp.) **Annex C** 05/08/2001 (genus listing for *Gonystylus* spp.)	**Appendix II** 23/06/2010 (genus listing for *Gonystylus* spp.) **Annex B** 20/12/2014 (genus listing for *Gonystylus* spp.) **Annotation** **#4** All parts and derivatives, except: **a)** seeds (including seedpods of Orchidaceae), spores and pollen (including pollinia). The exemption does not apply to seeds from Cactaceae spp. exported from Mexico, and to seeds from *Becca riophoenix madagascariensis* and *Neodypsis decaryi* exported from Madagascar; **b)** seedling or tissue cultures obtained in vitro, in solid or liquid media, transported in sterile containers; **c)** cut flowers of artificially propagated plants; **d)** fruits and parts and derivatives thereof of naturalized or artificially propagated plants of the genus *Vanilla* (Orchidaceae) and of the family Cactaceae; **e)** stems, flowers, and parts and derivatives thereof of naturalized or artificially propagated plants of the genera *Opuntia* subgenus *Opuntia* and Selenicereus (Cactaceae); and **f)** finished products of *Euphorbia antisyphilitica* packaged and ready for retail trade

Door frame

4418

Venetian blinds

Louvred doors

Paintbrushes

Clothes hangers

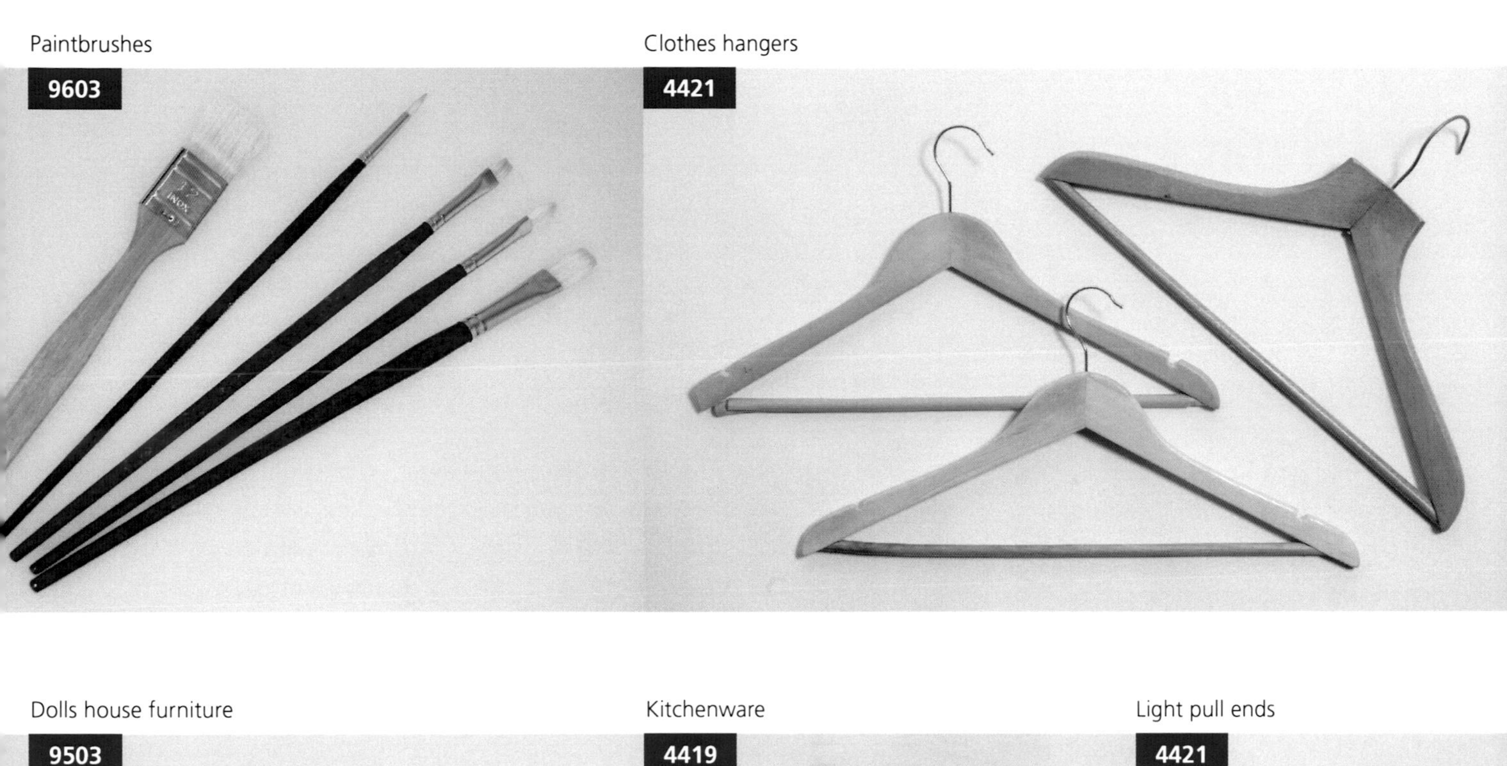

Dolls house furniture

Kitchenware

Light pull ends

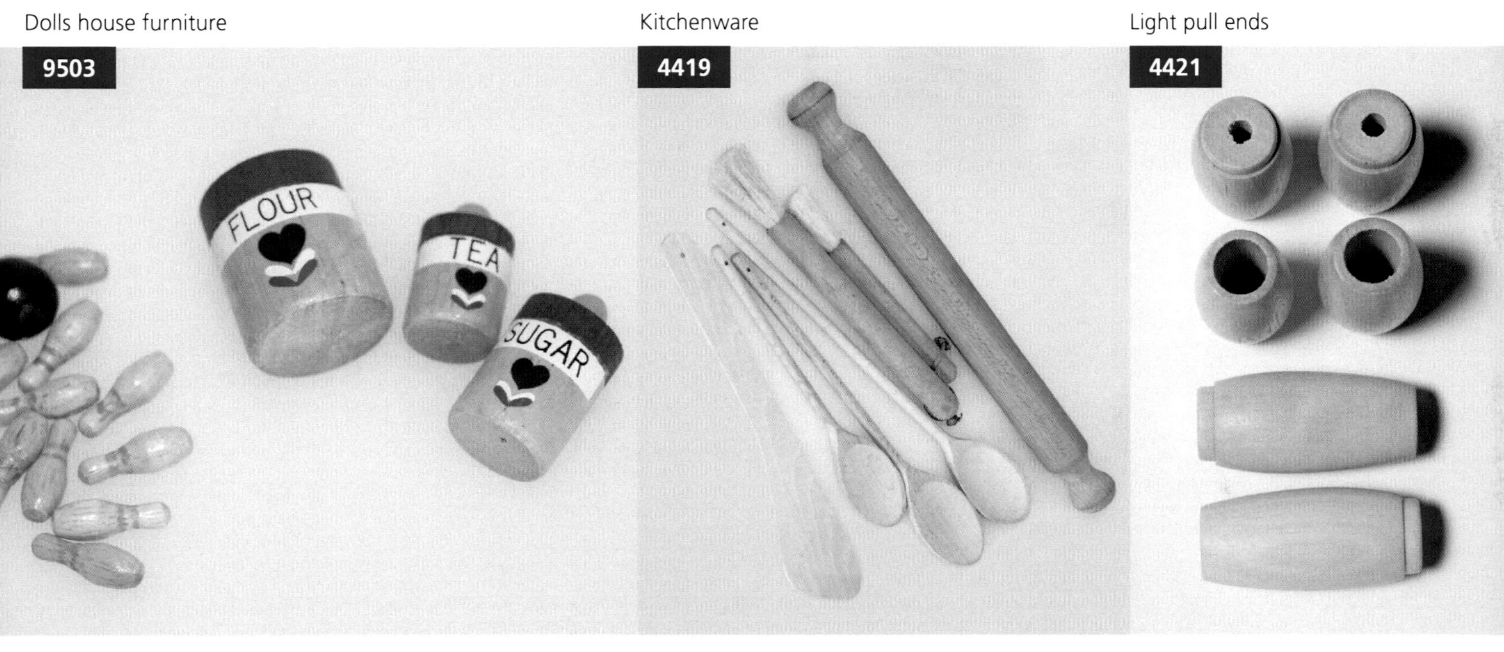

Plywood

Herbarium presses

Snooker cues

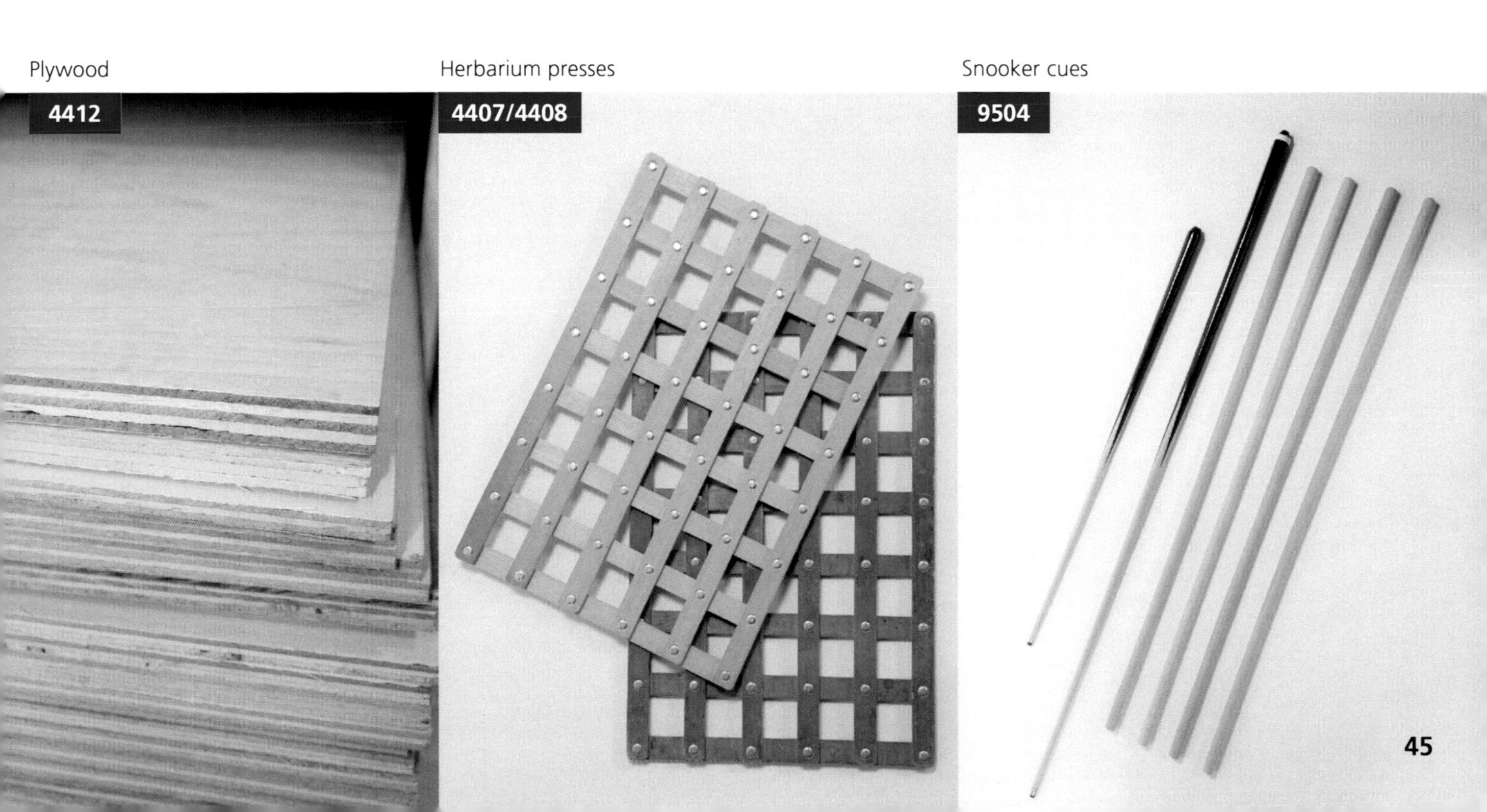

Guaiacum
Lignum vitae

Distribution

There are five species of tropical trees and shrubs in the genus *Guaiacum* which is listed under CITES. They are native to Central and South America and the Caribbean including Antigua and Barbuda, Bahamas, Barbados, Colombia, Cuba, Dominica, Dominican Republic, Grenada, Guadeloupe, Guatemala, Haiti, Honduras, Jamaica, Martinique, Mexico, Netherlands Antilles, Panama, Puerto Rico, St Vincent and the Grenadines, the United Kingdom Overseas Territories (Anguilla, British Virgin Islands, Monserrat, Turks and Caicos), the United States of America (Virgin islands) and Venezuela.

Uses

These species are popular in trade for mechanical devices (e.g. marine bearings and propeller shafts) as the wood has self-lubricating properties. They are also used in the manufacture of pulley sheaves, casters, bowling balls, and handles and sheaths. The species have medicinal uses and are sometimes cited as an ingredient in alcoholic drinks. The trade data does not always differentiate between the species, but the two major species in trade are *Guaiacum sanctum* and *G. coulteri*.

Trade

The vast majority of trade is in wild-sourced timber of *Guaiacum sanctum* exported from Mexico. Germany is the main importer and re-exporter, followed by Hong Kong and the USA. The other products in trade are extracts and powder with Germany and Switzerland as the largest importers and re-exporters.

The #2 annotation states that all parts and derivatives are regulated except seeds, pollen and finished products packaged and ready for retail trade. The definition of "finished products packaged and ready for retail trade" can be found in the Interpretation section to the Appendices and EU Annexes and in the CITES Glossary (**http://www.cites.org/eng/resources/terms/glossary.php**) as a "Product, shipped singly or in bulk, requiring no further processing, packaged, labelled for final use or the retail trade in a state fit for being sold to or used by the general public". This means that a product that requires no alteration or re-packaging and is ready for immediate use or sale is not regulated.

Plantations / artificial propagation

There are no large-scale commercial plantations of *Guaiacum* species, although plantation trials of *G.sanctum* have been carried out in Ghana. *Guaiacum officinale* and *G.sanctum* are planted as ornamental trees in the USA and other tropical areas. The majority of products in trade are wild in origin.

Pen blank bundles

Logs

CITES international trade suspensions, export quotas and reservations

There are no current CITES international trade suspensions, export quotas or reservations in place for this genus.

EU Decisions

There is a current EU opinion for *Guaiacum sanctum*.

See Species + for details **http://www.speciesplus.net/#/taxon_concepts?taxonomy=cites_eu&taxon_concept_query=Guaiacum&geo_entities_ids=&geo_entity_scope=cites&page=1**

SCIENTIFIC AND COMMON NAMES	DATE OF LISTING	CURRENT LISTING AND ANNOTATION
Scientific name and author: *Guaiacum angustifolia* Engelm. *Guaiacum coulteri* A.Gray *Guaiacum officinale* L. *Guaiacum sanctum* L. *Guaiacum unijugum* Brandegee	**Appendix II** 01/07/1975 (*G. sanctum*) 11/06/1992 (*G. officinale*) 13/02/2003 (genus listing for *Guaiacum* spp.)	**Appendix II** 13/09/2007 (genus listing for *Guaiacum* spp.)
Family: Zygophyllaceae	**Annex B** 01/06/1997 (*G. sanctum*) 01/06/1997 (*G. officinale*) 30/08/2003 (genus listing for *Guaiacum* spp.)	**Annex B** 20/12/2014 (genus listing for *Guaiacum* spp.)
Common names: English: sonora guaiacum, lignum-vitae, pockwood, wood of life, tree of life, ironwood French: bois de gaïac, bois de vie, bois saint, gaïac, gayac German: pockholz (as oil: guajaköl) Italian: guaiaco Portuguese: guaiaco, pau santo, lenha di guaiaco Spanish: guayacán, guajacum, leño de guayaco, madera de gaiac, palo de ropa, palo de vida, palo santo		**Annotation** #2 All parts and derivatives, except: a) seeds and pollen; and b) finished products packaged and ready for retail trade
CITES Standard Reference: The CoP has adopted a standard reference for this genus. It is *Lista de especies, nomenclatura y distribución en el genero Guaiacum.* Davila Aranda, P. & Schippmann, U. (2006): Medicinal Plant Conservation 12:50 http://www.cites.org/common/com/nc/tax_ref/Guaiacum.pdf		

Bearings – finished product

Pulley sheaves – finished product

Alcohol – finished product

Osyris lanceolata
African sandalwood

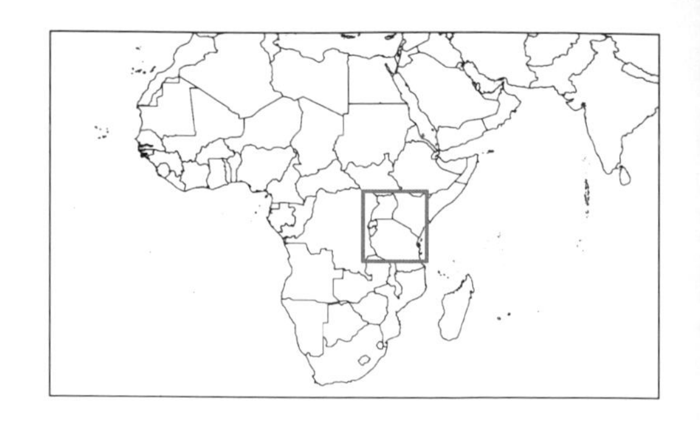

Distribution

There are between three and seven species in the genus *Osyris*. The taxonomy of this genus is unclear and requires more discussion to ensure other species are not included under the name *O. lanceolata*, but at present it is the only species in this genus listed under CITES. The original distribution of this small, evergreen, semi-parasitic shrub or small tree is unclear but is likely to have been Africa and parts of southern Europe. The current distribution is more widespread possibly due to the introduction of the species. It is now cited as being native to many sub-Saharan countries in Africa (Algeria to Ethiopia and south to South Africa), in restricted parts of southern Europe and in Asia (India to China). Outside of the main production areas for international trade (Kenya, Tanzania, South Sudan and Uganda), where it is now locally rare, the species seems to be widespread and common. Only the populations of Burundi, Ethiopia, Kenya, Rwanda, Uganda and Tanzania are regulated.

Uses

The heartwood and roots are distilled to produce an aromatic essential oil. It is increasingly being used as a substitute for other "sandalwood" producing genera (e.g. *Santalum* and *Pterocarpus* species) that are either banned from export or are in increasingly short supply. The tree is harvested in Kenya, Tanzania and other countries, semi-processed in Tanzania and the product is illegally exported through Mombasa, Kenya to Indonesia, India, South Africa, France, Germany and east Asian countries for the cosmetic and pharmaceutical industries.

Trade

The recent listing of this species in 2013 means there is no trade data available. The main products in international trade are essential oils or fragrances, cosmetics and toiletries containing the oil, handicrafts made from the timber and sawdust used in the manufacture of incense which may be compressed into cones or incense sticks. These products are available on the Internet.

The #2 annotation states that all parts and derivatives are regulated except seeds, pollen and finished products packaged and ready for retail trade. The definition of "finished products packaged and ready for retail trade" can be found in the Interpretation section to the Appendices and EU Annexes and in the CITES Glossary (**http://www.cites.org/eng/resources/terms/glossary.php**) as any "Product, shipped singly or in bulk, requiring no further processing, packaged,

Osyris lanceolata | Logs

labelled for final use or the retail trade in a state fit for being sold to or used by the general public". This means that a product that requires no alteration or re-packaging and is ready for immediate use or retail sale is not regulated.

Plantations / artificial propagation

There are no commercial plantations of this species meaning all trade is wild in origin.

CITES international trade suspensions, export quotas and reservations

There are no current CITES international trade suspensions or export quotas in place for this species.

There is a current reservation in place for this species.

EU Decisions

There are no current EU suspensions or opinions for this species.

See Species + for details **http://www.speciesplus.net/#/taxon_concepts/24847/legal**

SCIENTIFIC AND COMMON NAMES	DATE OF LISTING	CURRENT LISTING AND ANNOTATION
Scientific name and author: *Osyris lanceolata* Hochst. & Steud.	**Appendix II** 12/06/2013 (Populations of Burundi, Ethiopia, Kenya, Rwanda, Uganda and the United Republic of Tanzania)	**Appendix II** 12/06/2013 (Populations of Burundi, Ethiopia, Kenya, Rwanda, Uganda and the United Republic of Tanzania)
Family: Santalaceae		
Common names: **English:** East African sandalwood **German:** Ostafrikanisches sandelholz **Italian:** sandalo **Portugese:** sândalo-falso **Swahili:** msandali	**Annex B** 10/08/2013 (Populations of Burundi, Ethiopia, Kenya, Rwanda, Uganda and the United Republic of Tanzania)	**Annex B** 20/12/2014 (Populations of Burundi, Ethiopia, Kenya, Rwanda, Uganda and Tanzania) **Annotation** **#2** All parts and derivatives, except: a) seeds and pollen; and b) finished products packaged and ready for retail trade
CITES Standard Reference: The CoP has adopted a standard reference for generic names (*The Plant Book*, 2nd Edition, D. J. Mabberley, 1997, Cambridge University Press reprinted with corrections 1998). The Plant List is a useful source of information on scientific names and synonyms for this species (**http://www.theplantlist.org/tpl1.1/record/kew-2396402**). If you require more formal guidance contact the CITES Secretariat.		

Sawdust – unfinished

Essential oil – finished product

Logs

Pericopsis elata
Afrormosia

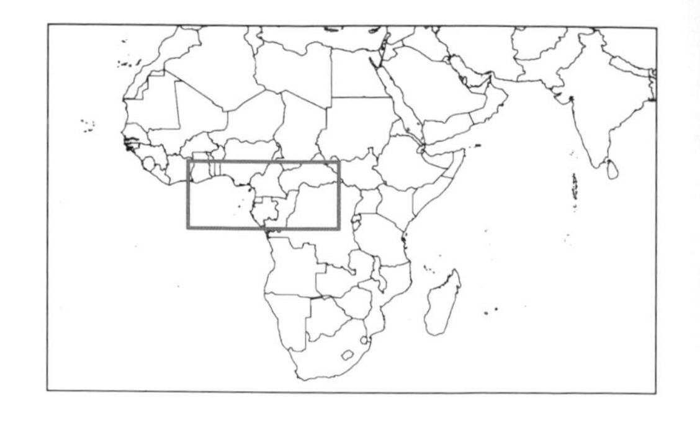

Distribution

Pericopsis elata is one of four species in the genus *Pericopsis* and the only one listed under CITES. It is native to the west and central African countries of Cameroon, Central African Republic, Cote d'Ivoire, Congo, Democratic Republic of Congo (DRC), Ghana and Nigeria.

Uses

Highly prized for its durable wood, the species has been used as a substitute for teak (*Tectonia* species) hence its common name of "African teak". The principal uses for the timber are for the manufacture of flooring, furniture, window/door frames, decorative veneers and boatbuilding.

Trade

The trade in Afrormosia is restricted to sawn timber, logs and veneer sheets. The main exporters are Cameroon and the DRC with Congo and Ghana exporting smaller amounts. The main importers are the EU, in particular Belgium and Italy, with additional high import volumes recorded by France, Germany, Portugal, mainland China, Taiwan, and Japan. The source of the timber imported by the EU is Cameroon and DRC. Discrepancies are evident in the trade data reported by exporters and importers. Imports into the EU recorded by exporters were not recorded by importers (Cyprus, Ireland, and the Netherlands). The main re-exporters of timber are the USA together with the EU (Belgium, France, Germany and Spain) and Turkey.

The #5 annotation means only logs, sawn wood and veneer sheets are regulated. Definitions of these timber terms are found in the CITES Glossary (**http://www.cites.org/eng/resources/terms/glossary.php**) and in Resolution Conf. 10.13 (Rev. CoP15) (**http://www.cites.org/eng/res/10/10-13R15.php**). Live plants and seeds are not regulated.

Plantations / artificial propagation

There are no large commercial plantations of this species but the species is grown in enrichment agroforestry systems and small plantations within its native range (e.g. Cameroon, Cote d'Ivoire, Ghana and DRC). The current trade will be wild in origin.

Pericopsis elata Logs Sawn wood

CITES international trade suspensions, export quotas and reservations

There are no current reservations in place for this species.

There is a current CITES international trade suspension in place for this species (see Notification 2014/039 **http://cites.org/sites/default/files/notif/E-Notif-2014-039_0.pdf**).

There are current export quotas in place for this species. Check Species + for details or check the annual export quotas on the CITES website (**http://www.cites.org/eng/resources/quotas/index.php**).

EU Decisions

There are a number of EU opinions for this species.

See Species + for details **http://www.speciesplus.net/#/taxon_concepts/18339/legal**

SCIENTIFIC AND COMMON NAMES	DATE OF LISTING	CURRENT LISTING AND ANNOTATION
Scientific name and author: *Pericopsis elata* (Harms) Meeuwen **Family**: Leguminosae **Common names:** **Cameroon:** assamela, ejen, obang **Cote d'Ivoire:** assamela **English:** African teak, golden afrormosia, satinwood, yellow satinwood **French:** teck d'Afrique **German:** afrormosia **Ghana:** kokrodua, awawai, mohole, Ghana asemela **Italian:** afrormosia **Portugese:** assamela, afrormosia, teca africana **Spanish:** afrormosia **CITES Standard Reference:** The CoP has adopted a standard reference for generic names (*The Plant Book*, 2nd Edition, D. J. Mabberley, 1997, Cambridge University Press reprinted with corrections 1998). The Plant List is a useful source of information on scientific names and synonyms for this species (**http://www.theplantlist.org/tpl1.1/record/ild-7160**). If you require more formal guidance contact the CITES Secretariat.	**Appendix II** 11/06/1992 **Annex B** 01/06/1997	**Appendix II** 13/09/2007 **Annex B** 20/12/2014 **Annotation** **#5** Logs, sawn wood and veneer sheets

Flooring

Logs

Veneer sheets

Pilgerodendron uviferum
Guaitecas cypress

Distribution

This slow-growing, evergreen conifer belongs to the Cypress family and is the only species in the genus *Pilgerodendron*. It is native to the temperate rainforests and sub-polar forests of Argentina and Chile.

Uses

The yellow-reddish timber of this species is decay resistant and has been heavily exploited for construction and the manufacture of carvings, house shingles, bridges, poles, fencing, boats and furniture.

Trade

This species is listed in Appendix I / Annex A and the international trade in wild-sourced specimens for commercial purposes is prohibited. Commercial trade in artificially propagated specimens is permitted. The CITES Trade Database indicates that carvings and seeds are the main products in trade along with much smaller quantities of live plants. The major exporter of carvings (source given as pre-Convention) is Argentina with the main importer being France. Wild-sourced seed exported from Chile for scientific purposes has been imported into the UK.

Plantations / artificial propagation

There are no known commercial plantations of this species. This species is planted as an ornamental tree and artificially propagated material (live plants and seed) is available from plant nurseries and for sale on the Internet.

CITES international trade suspensions, export quotas and reservations

There are no current CITES international trade suspensions, export quotas or reservations in place for this species.

EU Decisions

There are no current EU suspensions or opinions for this species.

See Species + for details **http://speciesplus.net/#/taxon_concepts/28405/legal**

Pilgerodendron uviferum forest

Illegal logging

SCIENTIFIC AND COMMON NAMES	DATE OF LISTING	CURRENT LISTING AND ANNOTATION
Scientific name and author: *Pilgerodendron uviferum* (D.Don) Florin	**Appendix I** 1/07/1975 **Annex A** 1/06/1997	**Appendix I** 1/07/1975 **Annex A** 20/12/2014 **Annotation** For Appendix I / Annex A species all parts and derivatives, live or dead, are regulated
Family: Cupressaceae		
Common names: English: Guaitecas cypress, Chilean cedar, Patagonian cypress, Patagonian pilgerodendron French: cèdre du Chile, cyprès du Chili German: Chilenische flußzeder Italian: larice del Cile, larice uvifero, cipresso del Cile Portugese: ciprés das Guaitecas Spanish: ciprés de las Guaitecas, cedro, ciprés Chileno		
CITES Standard Reference: The CoP has adopted a standard reference for generic names (*The Plant Book*, 2nd Edition, D. J. Mabberley, 1997, Cambridge University Press reprinted with corrections 1998). The Plant List is a useful source of information on scientific names and synonyms for this species (**http://www.theplantlist.org/tpl1.1/record/kew-2411520**). If you require more formal guidance contact the CITES Secretariat.		

House shingles

4418

Logs

4403

Logs

4403

Logs

4403

Pinus koraiensis
Korean pine

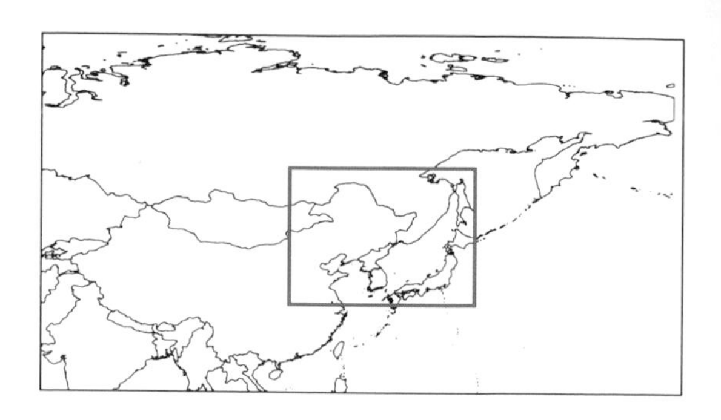

Distribution

This is the only species of pine in the genus *Pinus* (some 175 species) listed under CITES. The Korean pine is an evergreen conifer that can grow up to 30 metres in height. It is native to China, Japan, Democratic People's Republic of Korea, Republic of Korea and the Russian Federation.

Uses

The soft, straight grained and easily worked timber is used in the manufacture of telephone poles, railway sleepers, wooden bridges, boat building, flooring, plywood and veneers and can be chipped for particleboard manufacture, or pulped for the paper industry. It is also used in the manufacture of furniture, sports equipment and musical instruments. Resin extracted from the wood pulp is used to produce turpentine. The over-harvesting for its timber and edible seeds or "pine nuts" that are used in the food processing industry has resulted in a dramatic loss of habitat for the Siberian tiger (*Panthera tigris altaica*). There is a logging ban in the Russian Federation territories to assist the conservation of this species of which the pine forests are its key habitat. Live plants and seed of cultivated varieties of this species are found in international trade and are for sale on the Internet.

Trade

The recent listing of this species in 2010 means there is little trade data available. The CITES Trade Database indicates all trade is wild in origin and timber, carvings and sawn wood are the main products in trade. The major exporter of these products is the Russian Federation and the main importers are China (mainland and Hong Kong SAR) and Japan. This species is cited as the source of the majority of pine nuts imported into Europe and the USA.

The #5 annotation means only logs, sawn wood and veneer sheets are regulated. Definitions of these timber terms are found in the CITES Glossary (**http://www.cites.org/eng/resources/terms/glossary.php**) and in Resolution Conf. 10.13 (Rev. CoP15) (**http://www.cites.org/eng/res/10/10-13R15.php**). Live plants, seeds, pine nuts and finished products, are not regulated.

Plantations / artificial propagation

There are commercial plantations of this species in China and the Republic of Korea. Timber and pine nuts from these sources are in international trade.

Pinus koraiensis cones — Pine nuts — Plywood

CITES international trade suspensions, export quotas and reservations

There are no current CITES international trade suspensions, export quotas or reservations in place for this species

EU Decisions

There are no current EU suspensions or opinions for this species.

See Species + **http://speciesplus.net/#/taxon_concepts/18210/legal**

<table>
<tr><th>SCIENTIFIC AND COMMON NAMES</th><th>DATE OF LISTING</th><th>CURRENT LISTING AND ANNOTATION</th></tr>
<tr><td>Scientific name and author: Pinus koraiensis Siebold & Zucc</td><td rowspan="4">Appendix III
14/10/2010

Annex C
14/02/2012</td><td rowspan="4">Appendix III
14/10/2010

Annex C
20/12/2014

Annotation
#5 Logs, sawn wood and veneer sheets.</td></tr>
<tr><td>Family: Pinaceae</td></tr>
<tr><td>Common names:
Chinese: 红松 hong song
English: Korean pine, Korean nut pine, Korean white pine, Korean cedar, Korean pine
German: Korea-kiefer
Italian: pino coreano
Japanese: チョウセンゴヨウ chosen-goyo, chosen-matsu
Korean: 잣나무
Portugese: pinho da Coreia, pinheiro da Coreia
Russian: Корейский кедр
Swedish: koreatall</td></tr>
<tr><td>CITES Standard Reference: The CoP has adopted a standard reference for generic names (The Plant Book, 2nd Edition, D. J. Mabberley, 1997, Cambridge University Press reprinted with corrections 1998). The Plant List is a useful source of information on scientific names and synonyms for this species (http://www.theplantlist.org/tpl1.1/record/kew-2561659).
If you require more formal guidance contact the CITES Secretariat.</td></tr>
</table>

Telegraph poles

Pine logs

Platymiscium pleiostachyum
Cristóbal

Distribution

Platymiscium pleiostachyum is one of 35 species in the genus *Platymiscium*. It is the only species from this genus listed under CITES and is native to Costa Rica, El Salvador and Nicaragua.

Uses

The wood of this species is an attractive red to reddish brown in colour and is used in the manufacture of panelling, flooring, cabinet making, furniture and musical instruments (drums).

Trade

According to the CITES Trade Database there are very low levels of international trade in this species. Only two records are indicated – one carving (reported as artificially propagated) was exported by Costa Rica in 2003 to the USA and one unit of timber (reported as wild-sourced for scientific reasons) from El Salvador in 2006, also to the USA.

The #4 annotation means that all parts and derivatives, live or dead, are regulated except for seeds, tissue cultured plants in sterile containers and cut flowers from artificially propagated flowers.

Plantations / artificial propagation

There are no known commercial plantations of this species and any trade will be wild in origin.

CITES international trade suspensions, export quotas and reservations

There are no current CITES international trade suspensions, export quotas or reservations in place for this species.

EU Decisions

There are no current EU suspensions or opinions for this species.

See Species + for details **http://speciesplus.net/#/taxon_concepts/28731/legal**

Veneer sheets

Wood panelling

SCIENTIFIC AND COMMON NAMES	DATE OF LISTING	CURRENT LISTING AND ANNOTATION
Scientific name and author: *Platymiscium pleiostachyum* Donn.Sm	**Appendix I** 1/07/1975	**Appendix II** 23/06/2010
Family: Leguminosae	**Annex B** 1/06/1997	**Annex B** 20/12/2014
Common names: **English:** platymiscium **German:** granadillo, cristobal **Italian:** granadillo **Portugese:** macacaúba **Spanish:** ambar, cachimbo, coyote, cristóbal, granadillo, guayacan trebol, jacaranda do brejo, macacauba, macawood, ñambar		**Annotation** **#4** All parts and derivatives, except: **a)** seeds (including seedpods of Orchidaceae), spores and pollen (including pollinia). The exemption does not apply to seeds from Cactaceae spp. exported from Mexico, and to seeds from *Beccariophoenix madagascariensis* and *Neodypsis decaryi* exported from Madagascar; **b)** seedling or tissue cultures obtained in vitro, in solid or liquid media, transported in sterile containers; **c)** cut flowers of artificially propagated plants; **d)** fruits and parts and derivatives thereof of naturalized or artificially propagated plants of the genus *Vanilla* (Orchidaceae) and of the family Cactaceae; **e)** stems, flowers, and parts and derivatives thereof of naturalized or artificially propagated plants of the genera *Opuntia* subgenus *Opuntia* and Selenicereus (Cactaceae); and **f)** finished products of *Euphorbia antisyphilitica* packaged and ready for retail trade
CITES Standard Reference: The CoP has adopted a standard reference for generic names (*The Plant Book*, 2nd Edition, D. J. Mabberley, 1997, Cambridge University Press reprinted with corrections 1998). The Plant List is a useful source of information on scientific names and synonyms for this species (**http://www.theplantlist.org/tpl1.1/record/ild-15750**). If you require more formal guidance contact the CITES Secretariat.		

Wood flooring

4409

Podocarpus neriifolius
Podocarp

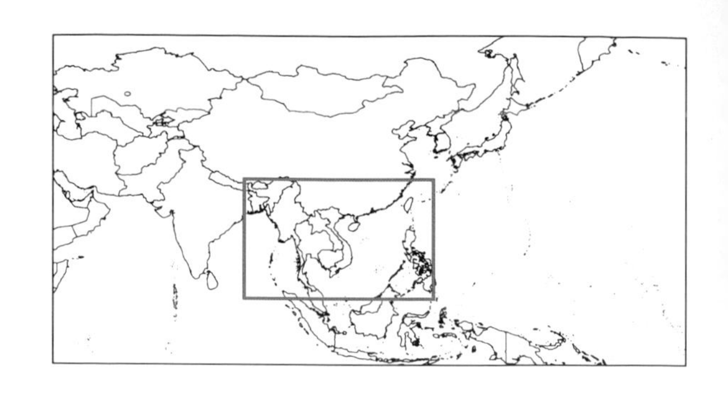

Distribution

This is one of two species in the genus *Podocarpus* listed under CITES. This conifer is native to Brunei Darussalam, Cambodia, China, Fiji, India, Indonesia, Lao People's Democratic Republic, Malaysia (Peninsular Malaysia, Sabah, Sarawak), Myanmar, Nepal, Papua New Guinea, the Philippines, the Solomon Islands, Thailand, and Viet Nam.

Uses

The timber is used for house building, carpentry and for the manufacture of paper, oars and masts of sailing vessels. Higher grades of timber are used for veneer, furniture making, cabinet making, household utensils, musical instruments and carvings.

Trade

The CITES Trade Database indicates very low levels of international trade in this species. Only 500 live plants were in trade in the period 2004–2014. The main exporters of live plants are Myanmar, Malaysia and Denmark. The main importers are China and Switzerland.

The #1 annotation means all parts and derivatives, live or dead, are regulated apart from seeds, tissue cultured plants in sterile containers and cut flowers from artificially propagated plants.

Plantations / artificial propagation

There are no known commercial plantations of this species. Live plants are artificially propagated and in trade for use in the horticultural industry.

CITES international trade suspensions, export quotas and reservations

There are no current CITES international trade suspensions, export quotas or reservations in place for this species.

Podocarpus neriifolius | Oars | Kitchen utensils

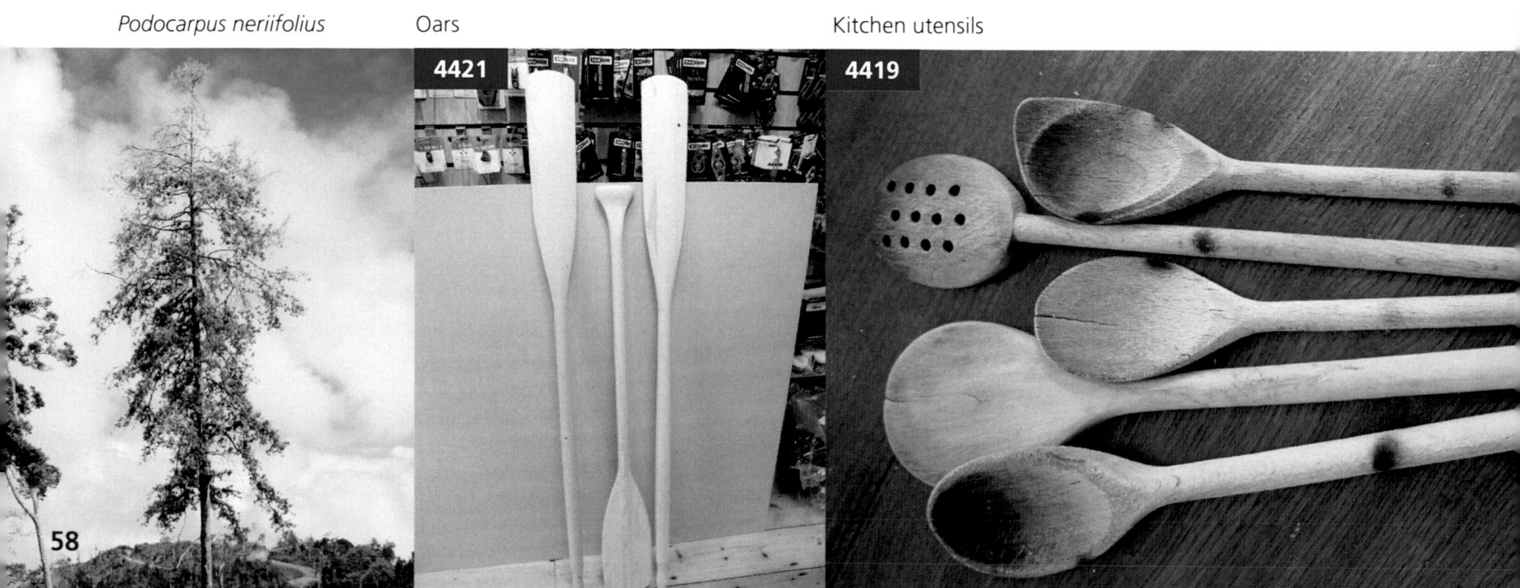

There is a trade suspension in place for the export of all wild specimens of Appendix I, II and III species from India (see Notification 1999/039 **http://cites.org/sites/default/files/eng/notif/1999/039.shtml**). However, export permits will be issued for cultivated varieties of plant species included in Appendices I and II.

EU Decisions

There are no current EU suspensions or opinions for this species

See Species + for details **http://speciesplus.net/#/taxon_concepts/27490/legal**

SCIENTIFIC AND COMMON NAMES	DATE OF LISTING	CURRENT LISTING AND ANNOTATION
Scientific name and author: *Podocarpus neriifolius* D.Don	**Appendix III** 16/11/1975 **Annex C** 01/06/1997	**Appendix III** 23/06/2010 **Annex C** 20/12/2014 **Annotation** **#1** All parts and derivatives, except: **a)** seeds, spores and pollen (including pollinia); **b)** seedling or tissue cultures obtained in vitro, in solid or liquid media, transported in sterile containers; **c)** cut flowers of artificially propagated plants; and **d)** fruits, and parts and derivatives thereof, of artificially propagated plants of the genus *Vanilla*
Family: Podocarpaceae		
Common names: **Chinese:** bai ri qing **English:** brown pine, black pine podocarp **German:** podo, maniu **Lao:** ka dong **Malay:** podo bukit **Portuguese:** pinho bravo **Vietnamese:** thông tre lá dài **Local:** amanu,bukiti, jati bukit		
CITES Standard Reference: The CoP has adopted a standard reference for generic names (*The Plant Book*, 2nd Edition, D. J. Mabberley, 1997, Cambridge University Press reprinted with corrections 1998). The Plant List is a useful source of information on scientific names and synonyms for this species **http://www.theplantlist.org/tpl1.1/record/kew-2567549** If you require more formal guidance contact the CITES Secretariat.		

Ship mast

Table

Podocarpus parlatorei
Parlatore's podocarp

Distribution

This is one of two species of the genus *Podocarpus* listed under CITES. This conifer is native to Bolivia and Argentina.

Uses

The timber is light in colour, lightweight and relatively soft. It is used to make fence posts, utensils, houses and pencils.

Trade

This species is listed in Appendix I / Annex A and the international trade in wild-sourced specimens for commercial purposes is prohibited. Commercial trade in artificially propagated specimens is permitted. The CITES Trade Database indicates little international trade with only two trade records being shown. These were for the export of leaves and timber (wild-sourced for scientific purposes) from Bolivia to Argentina.

Plantations / artificial propagation

There are no known commercial plantations of this species and any trade in timber will be wild in origin.

CITES international trade suspensions, export quotas and reservations

There are no current CITES international trade suspensions, export quotas or reservations in place for this species.

EU Decisions

There are no current EU suspensions or opinions for this species.

See Species + for details **http://www.speciesplus.net/#/taxon_concepts/22858/legal**

Podocarpus parlatorei

Fenceposts

SCIENTIFIC AND COMMON NAMES	DATE OF LISTING	CURRENT LISTING AND ANNOTATION
Scientific name and author: *Podocarpus parlatorei* Pilg.	**Appendix I** 1/07/1975 **Annex A** 1/06/1997	**Appendix I** 1/07/1975 **Annex A** 20/12/2014 **Annotation** As an Appendix I / Annex A species all parts and derivatives, live or dead, are regulated
Family: Podocarpaceae		
Common names: **English:** Parlatore's podocarp, yellow wood, brown pine, black pine **German:** podo, maniu **Portugese:** pinho do monte **Spanish:** monteromero, pino blanco, pino del cerro, pino montano		
CITES Standard Reference: The CoP has adopted a standard reference for generic names (*The Plant Book*, 2nd Edition, D. J. Mabberley, 1997, Cambridge University Press reprinted with corrections 1998). The Plant List is a useful source of information on scientific names and synonyms for this species (http://www.theplantlist.org/tpl1.1/record/kew-2567526). If you require more formal guidance contact the CITES Secretariat.		

Pencils

9609

Brushes

4417

Prunus africana
African cherry

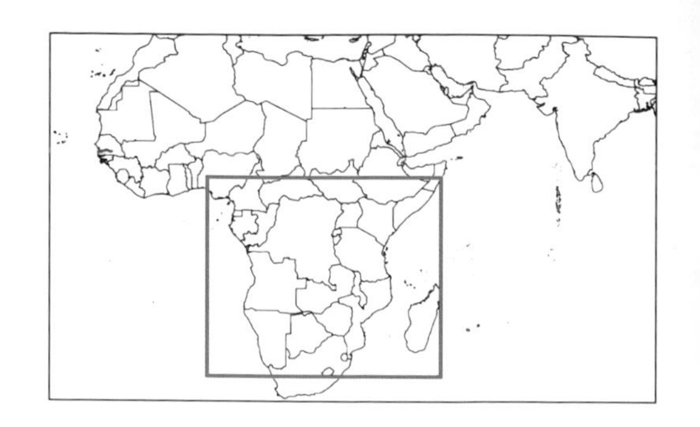

Distribution

The African cherry, a wild relative of plums, cherries and almonds, is the only species in the genus *Prunus* listed under CITES. It is native to Angola, Birundi, Cameroon, Comoros, Congo, Democratic Republic of Congo, Equitorial Guinea, Ethiopia, Kenya, Madagascar, Malawi, Mozambique, Nigeria, Rwanda, Sao Tome and Principe, South Africa, Sudan (prior to secession to southern Sudan), Swaziland, Uganda, United Republic of Tanzania, Zambia and Zimbabwe.

Uses

Extracts from its pungent bark are traded internationally as an herbal remedy for benign prostate hyperplasia. There are at least 40 brand-name products using *Prunus africana* bark extract as an ingredient. High street herbal products containing *Pygeum africanum*, which is a synonym of *P. africana*, are readily available over the counter or for sale on the Internet.

Trade

The main products in international trade are unprocessed dried bark and processed medicines. The traditional importers have been European companies, but there are emerging Asian markets in China and India. The major exporters of bark are Cameroon, Uganda and DRC, and the main importer is France, with Spain and Italy importing lower volumes. The main re-exporters of extracts are France and Spain.

The #4 annotation means that all parts and derivatives are regulated except seeds, tissue cultured plants in sterile containers and cut flowers from artificially propagated plants. This means that crude (unprocessed bark, powders) or processed products (pills) whether finished and ready for retail trade or not are regulated.

Plantations / artificial propagation

There are no commercial plantations of this species and the majority of harvested bark comes from wild trees.

Small scale propagation schemes and trees planted as part of agroforestry or small agricultural farms systems, particularly in western Cameroon, have been ongoing since the 1970s. Bark from these sources is unlikely to be in international trade.

CITES international trade suspensions, export quotas and reservations

There are no current reservations in place for this species.

Harvesting bark

Artificially propagated seedlings

Bark harvesting

There are current CITES international trade suspensions in place for this species. See CITES Notification No. 2014/039 **(http://cites.org/sites/default/files/notif/E-Notif-2014-039_0.pdf).**

There is a current export quota for this species. Check Species + for details or check the annual export quotas on the CITES website (**http://www.cites.org/eng/resources/quotas/index.php**).

EU decisions

There are a number of current EU opinions for this species.

See Species + for details **http://www.speciesplus.net/#/taxon_concepts/22086/legal**

SCIENTIFIC AND COMMON NAMES	DATE OF LISTING	CURRENT LISTING AND ANNOTATION
Scientific name and author: *Prunus africana* (Hook.f.) Kalkman **Family**: Rosaceae	**Appendix II** 16/02/1995 **Annex B** 01/06/1997	**Appendix II** 23/06/2010 **Annex B** 20/12/2014
Common names: **English:** African cherry, African stinkwood, bitter almond, blackwood, ironwood, *Pygeum africanum*, red stinkwood, red strinkhout **Cameroon:** alumty, iluo, kirah, vla, wotangue **French:** teck d'Afrique **German:** Afrikanisches stinkholz **Kenya:** muiri **Madagascar:** kotofy, pesopeso **Portugese:** cerejeira africana **Spanish:** ciruelo africano, mueri **CITES Standard Reference:** The CoP has adopted a standard reference for generic names (*The Plant Book*, 2nd Edition, D. J. Mabberley, 1997, Cambridge University Press reprinted with corrections 1998). The Plant List is a useful source of information on scientific names and synonyms for this species (**http://www.theplantlist.org/tpl1.1/record/rjp-25340**). If you require more formal guidance contact the CITES Secretariat.		**Annotation** **#4** All parts and derivatives, except: **a)** seeds (including seedpods of Orchidaceae), spores and pollen (including pollinia). The exemption does not apply to seeds from Cactaceae spp. exported from Mexico, and to seeds from *Beccariophoenix madagascariensis* and *Neodypsis decaryi* exported from Madagascar; b) seedling or tissue cultures obtained in vitro, in solid or liquid media, transported in sterile containers; c) cut flowers of artificially propagated plants; d) fruits and parts and derivatives thereof of naturalized or artificially propagated plants of the genus *Vanilla* (Orchidaceae) and of the family Cactaceae; e) stems, flowers, and parts and derivatives thereof of naturalized or artificially propagated plants of the genera *Opuntia* subgenus *Opuntia* and *Selenicereus* (Cactaceae); and f) finished products of *Euphorbia antisyphilitica* packaged and ready for retail trade

Unprocessed bark

1211

Pills

3304

Unprocessed bark

1211

Pterocarpus santalinus
Red sandalwood

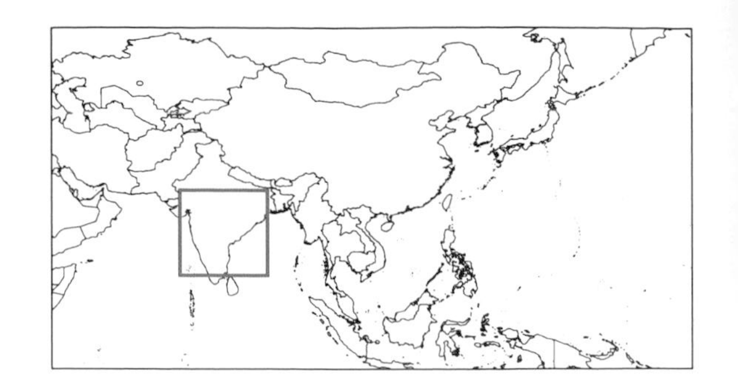

Distribution

Of the 66 or so species in the genus *Pterocarpus* this is the only species listed under CITES. This species is endemic to India but has been introduced on a small scale to Sri Lanka and Pakistan.

Uses

This species is renowned for its claret-red heartwood. The wood is used to make traditional instruments ('shamisen') and name seals ('hankos') in Japan and it yields a red pigment, santalin, used as a dye and colouring agent in cosmetics, pharmaceutical preparations and foodstuffs. This species is among the timber species classified under China's National Hongmu Standard for use in the manufacture of luxury deep red coloured Hongmu furniture. China is the only country that has a specific customs code for Hongmu species – 44039930. This species is traded under the common name of "sandalwood" and may be confused with other sandalwood producing tree species and genera, which are CITES-listed (*Osyris lanceolata*) and non CITES-listed (*Amyris balsamifera*, *Baphia nitida, Fusanus spicatus* and *Santalum* species).

Trade

The main products in trade are timber, wood chips and powder. There should be no commercial trade in wild specimens of this species due to a zero export quota set by India for this and other native species. However, extensive seizures and an auction of this illegally harvested timber means wild-sourced specimens are in trade. The main exporter of timber is India and the main importer of timber is China. The main re-exporter is Singapore. Virtually all trade in powder is in Europe, (Germany and Switzerland), re-exported between each other.

The #7 annotation means that only logs, woodchips, powder and extracts are regulated. The definition of wood chips, extract and powder can be found in the Interpretation section of the Appendices and EU Annexes or in the CITES Glossary (**http://www.cites.org/eng/resources/terms/glossary.php**).

Plantations / artificial propagation

There are commercial plantations of this species in India (approximately 3,000 hectares mainly in the states of Andhra Pradesh and Tamil Nadu) but it is not clear whether they are producing sufficient quantities for commercial purposes and export. Cultivation on farm land using this species is also carried out but no comprehensive inventories of this type of production have been made. Material from other sandalwood producing genera grown in plantations may be in trade and confused with *Pterocarpus santalinus* e.g. *Santalum album* (known as Indian sandalwood) is grown in plantations mainly in India and Australia.

Pterocarpus santalinus Name seals – Hankos Shamisen

CITES international trade suspensions, export quotas and reservations

There are no CITES international trade suspensions or reservations in place for this species

There is a current zero export quota in place for this species from India. However, a footnote states that "India will authorize the export of specimens of any type, from 310 metric tonnes of wood per year from artificially propagated source (Source "A") and a one-time export of specimens of any type, from 11,806 metric tones of wood from confiscated or seized source (Source "I"). Check Species + for details or check the annual export quotas on the CITES website (**http://www.cites.org/eng/resources/quotas/index.php**).

There is a trade suspension in place for the export of all wild specimens of Appendix I, II and III species from India (see Notification 1999/039 **http://cites.org/sites/default/files/eng/notif/1999/039.shtml**). However, export permits will be issued for cultivated varieties of plant species included in Appendices I and II.

EU decisions

There is a current EU negative opinion for this species from India.

See Species + for details **http://www.speciesplus.net/#/taxon_concepts/28106/legal**

SCIENTIFIC AND COMMON NAMES	DATE OF LISTING	CURRENT LISTING AND ANNOTATION
Scientific name and author: *Pterocarpus santalinus* L.f.	**Appendix II** 16/02/1995	**Appendix II** 13/09/2007
Family: Leguminosae	**Annex B** 01/06/1997	**Annex B** 20/12/2014
Common names: English: agaru, red sanders, red sandalwood, sandalwood, saunderswood French: bois de santal rouge, santal rouge German: rotes sandelholz Italian: sandalo rosso Portugese: sândalo vermelho Spanish: leno de sándalo rojo, sándalo rojo		**Annotation** #7 Logs, woodchips, powder and extracts
CITES Standard Reference: The CoP has adopted a standard reference for generic names (*The Plant Book*, 2nd Edition, D. J. Mabberley, 1997, Cambridge University Press reprinted with corrections 1998). The Plant List is a useful source of information on scientific names and synonyms for this species (**http://www.theplantlist.org/tpl1.1/record/ild-32307**). If you require more formal guidance contact the CITES Secretariat.		

Essential oil

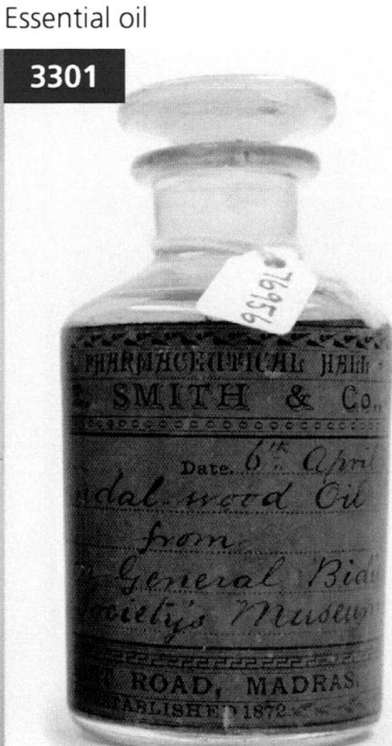

Hongmu chair carving

Quercus mongolica
Mongolian oak

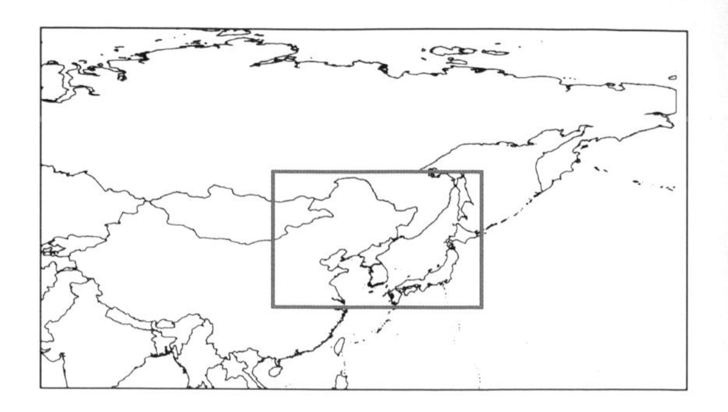

Distribution

This is the only one of some 600 species of oak regulated under CITES. This species is native to China, Japan, Republic of Korea, Mongolia and the Russian Federation (Sakalin Island).

Uses

The hard, heavy, straight-grained timber is durable and decay resistant and is used in the manufacture of mine and telegraph poles, stakes and sports equipment, boats, vehicles, agricultural tools, bridges and coopery.

Trade

The recent listing of this species in 2014 means there is no trade data available.

The #5 annotation means only logs, sawn wood and veneer sheets are regulated. Definitions of these timber terms are found in the CITES Glossary (**http://www.cites.org/eng/resources/terms/glossary.php**) and in Resolution Conf. 10.13 (Rev. CoP15) (**http://www.cites.org/eng/res/10/10-13R15.php**). Live plants and seeds are not regulated.

Plantations / artificial propagation

There are plantations of this species in China, but their production levels are unknown. The species is used in afforestation projects in Japan and was introduced into Europe in 1879 so small quantities of timber from such trees may be in trade. Artificially propagated live plants and seed are available from horticultural nurseries worldwide and may be found in international trade, although there may be plant health restrictions on the import of oak species into certain countries.

CITES international trade suspensions, export quotas and reservations

There are no current CITES international trade suspensions, export quotas or reservations in place for this species.

Quercus mongolica | Telegraph poles | Logs

EU Decisions

There are no current EU suspensions or opinions for this species.

See Species + for details **http://www.speciesplus.net/#/taxon_concepts/65569/legal**

SCIENTIFIC AND COMMON NAMES	DATE OF LISTING	CURRENT LISTING AND ANNOTATION
Scientific name and author: *Quercus mongolica* Fisch. ex Ledeb **Family:** Fagaceae **Common names:** **English:** Mongolian oak **German:** Mongolische eiche **Italian:** quercia mongola **Portugese:** carvalho da Mongólia **CITES Standard Reference:** The CoP has adopted a standard reference for generic names (*The Plant Book*, 2nd Edition, D. J. Mabberley, 1997, Cambridge University Press reprinted with corrections 1998). The Plant List is a useful source of information on scientific names and synonyms for this species (**http://www.theplantlist.org/tpl1.1/record/kew-173983**). If you require more formal guidance contact the CITES Secretariat.	**Appendix III** 24/06/2014 **Annex C** 20/12/2014	**Appendix III** 24/06/2014 **Annex C** 20/12/2014 **Annotation** **#5** Logs, sawn wood and veneer sheets.

Tool handles

Barrels

Swietenia
Mahoganies

Distribution

There are three species in the genus *Swietenia* (*S. macrophylla, S. humilis* and *S. mahagoni*) all of which are listed under CITES. These species are native to the Neotropics (Central and South America and the Caribbean) including Antigua and Barbuda, Bahamas, Barbados, Belize, Bolivia, Brazil, Colombia, Costa Rica, Cuba, Dominica, Dominican Republic, Ecuador, El Salvador, Grenada, French Guiana, Guadeloupe, Guatemala, Guyana, Honduras, Jamaica, Martinique, Mexico, Nicaragua, Panama, Peru, St Kitts and Nevis, St Lucia, St Vincent and the Grenadines, Trinidad and Tobago, the United Kingdom Overseas Territories (Anguilla, Cayman Islands, Turks and Caicos) and Venezuela. See Species + for individual species distribution: **http://www.speciesplus.net/#/taxon_concepts?taxonomy=cites_eu&taxon_concept_query=Swietenia%20&geo_entities_ids=&geo_entity_scope=cites&page=1**.

Uses

The species *Swietenia mahagoni* is considered commercially extinct but was popular in the manufacture of musical instruments, in particular guitars, and is used as veneers and in carpentry in general. The timber from *S. humilis* is used for veneers, restoration and musical instruments. Timber from *S. macrophylla* is in trade for the manufacture of furniture, panelling and musical instruments (e.g. guitars, ukuleles) and sports equipment (e.g. billiard tables). The common name "mahogany" may also refer to non-CITES listed timber species or genera including *Khaya* and *Entandrophragma* species (known as African mahogany), *Shorea* (Meranti, Balau, or Lauan), New Zealand mahogany or kohekohe (*Dysoxylum spectabile*), *Toona sinensis* (Chinese mahogany), *Toona sureni* (Indonesian mahogany), *Toona ciliata* (Indian mahogany), *Toona calantas* (Philippine mahogany), *Melia azedarach* (Chinaberry), *Guarea* (Pink Mahogany or Bosse), Chittagong (also known as Indian Mahogany), *Chukrasia velutina* and *Carapa guianensis* (Crabwood).

Trade

Big leaf mahogany (*Swietenia macrophylla*) is the most heavily exploited and common *Swietenia* species in international trade. Trade data for *S. humilis* indicates that Guatemala is the main exporter with the USA as the main importer. For *S. mahagoni*, which is considered commercially extinct, trade is mainly in pre-Convention timber/specimens and antiques. The major products of *S. macrophylla* seen in trade are timber (sawn wood and veneer) and lower volumes of carvings,

Swietenia macrophylla | Big-leaf mahogany guitars | Logs

logs and plywood. The CITES Trade Database indicates that the main exporter is Peru and, particularly after 2006, Bolivia, Guatemala and Mexico. Exports were also reported by Brazil, Belize, Colombia, Ecuador and Nicaragua. The USA is the main importer for mahogany species along with the Dominican Republic. The EU (Denmark, Germany and Spain) is a minor importer, the source given as Brazil, Peru and Nicaragua. The EU (Germany, Italy and Spain) also re-exports timber and veneer, mainly to the USA.

Sunken logs of *Swietenia macrophylla* are salvaged from rivers, such as those in Honduras and Belize, where there has been a history of logging this species since the early 20th century.

The #4 annotation (applicable to *Swietenia humilis*) means all parts and derivatives are regulated apart from seeds, pollen, tissue cultured plants in sterile containers and cut flowers from artificially propagated plants. The #6 annotation (applicable to *S. macrophylla*) means only logs, sawn wood, veneer sheets and plywood are regulated and the #5 annotation (applicable to *S. mahagoni*) means only logs, sawn wood and veneer sheets are regulated. Definitions of these timber terms are found in the CITES Glossary (**http://www.cites.org/eng/resources/terms/glossary.php**) and in Resolution Conf. 10.13 (Rev. CoP15) (**http://www.cites.org/eng/res/10/10-13R15.php**). Under the # 6 and # 5 annotations finished products, such as a finished guitar or a finished table with a veneer of mahogany, are not regulated. Any timber of *S. macrophylla* grown and exported from outside of the Neotropics is not regulated.

Plantations / artificial propagation

Commercial plantations of *Swietenia macrophylla* are found in China, Fiji, Bangladesh, India, Indonesia, the Philippines and Sri Lanka and timber is in international trade from some of these countries, in particular Fiji. Small amounts of timber or artefacts made from specimens of all three species, which have been planted as ornamentals both within and outside of their natural range, may be in trade.

CITES international trade suspensions, export quotas and reservations

There are no current CITES international trade suspensions or reservations in place for these species.

There are a number of current export quotas in place for *Swietenia macrophylla.* Check Species + for details or check annual export quotas on the CITES website (**http://www.cites.org/eng/resources/quotas/index.php**).

EU Decisions

There are a number of current EU opinions for *S.macrophylla.*

See Species + for details **http://www.speciesplus.net/#/taxon_concepts?taxonomy=cites_eu&taxon_concept_query=Swietenia%20&geo_entities_ids=&geo_entity_scope=cites&page=1**

Sawn timber

Big-leaf mahogany candlesticks

Sawn timber

SCIENTIFIC AND COMMON NAMES	DATE OF LISTING	CURRENT LISTING AND ANNOTATION
Scientific names and authors: *Swietenia humilis* Zucc. *Swietenia macrophylla* King *Swietenia mahagoni* (L.) Jacq.	*S. humilis* **Appendix II:** 01/07/1975 **Annex B:** 01/06/1997 *S. macrophylla* **Appendix III :** 16/11/1995 **Annex C:** 01/06/1997 *S. mahagoni* **Appendix II:** 11/06/1992 **Annex B:** 01/06/1997	*S. humilis* **Appendix II**: 23/06/2010 **Annex B**: 20/12/2014 *S. macrophylla* **Appendix II:** 15/11/2003 **Annex B:** 20/12/2014 *S. mahagoni* **Appendix II:** 18/09/1997 **Annex B:** 20/12/2014 **Annotation:** *S. humilis* **#4** All parts and derivatives, except: a) seeds (including seedpods of Orchidaceae), spores and pollen (including pollinia). The exemption does not apply to seeds from Cactaceae spp. exported from Mexico, and to seeds from *Beccariophoenix madagascariensis* and *Neodypsis decaryi* exported from Madagascar; b) seedling or tissue cultures obtained in vitro, in solid or liquid media, transported in sterile containers; c) cut flowers of artificially propagated plants; d) fruits and parts and derivatives thereof of naturalized or artificially propagated plants of the genus *Vanilla* (Orchidaceae) and of the family Cactaceae; e) stems, flowers, and parts and derivatives thereof of naturalized or artificially propagated plants of the genera *Opuntia* subgenus *Opuntia* and Selenicereus (Cactaceae); and f) finished products of *Euphorbia antisyphilitica* packaged and ready for retail trade *S. macrophylla* (Populations of the Neotropics) **#6** Logs, sawn wood, veneer sheets and plywood *S. mahagoni* **#5** Logs, sawn wood and veneer sheets
Family: Meliaceae		
Common names: English: big-leaf mahogany, bigleafed mahogany, Brazilian mahogany (*S. macrophylla*), West Indian or Cuban mahogany, Honduran mahogany, American mahogany (*S. mahagoni*), Mexican or Honduran mahogany (*S. humilis*) French: acajou d'Amérique, swiétenie, acajou d'Amérique Centrale, acajou du Honduras, (*S. macrophylla*), acajou du Mexique, acajou de la côte du Pacifique (*S. humilis*), acajou à meubles, mahogani petites feuilles, acajou des Antilles, acajou ronceux (*S. mahagoni*) German: Amerikanisches mahagoni, mogno Italian: mogano Americano Portugese: mogno Spanish: caoba, mara, aguano or ahuano (*S. macrophylla*), cobano, caoba de Honduras, caoba de Pacifica (*S. humilis*), caoba Española, acajou de Santo Domingo, caoba Americana (*S. mahagoni*)		
CITES Standard Reference: The CoP has adopted a standard reference for this genus (*The Plant Book*, 2nd edition, D. J. Mabberley, 1997, Cambridge University Press reprinted with corrections 1998). The Plant List is also a useful source of information on scientific names and synonyms **http://www.theplantlist.org/tpl1.1/search?q=Swietenia** If you require more formal guidance contact the CITES Secretariat.		

Measuring sawn wood for export

Taxus
Yews

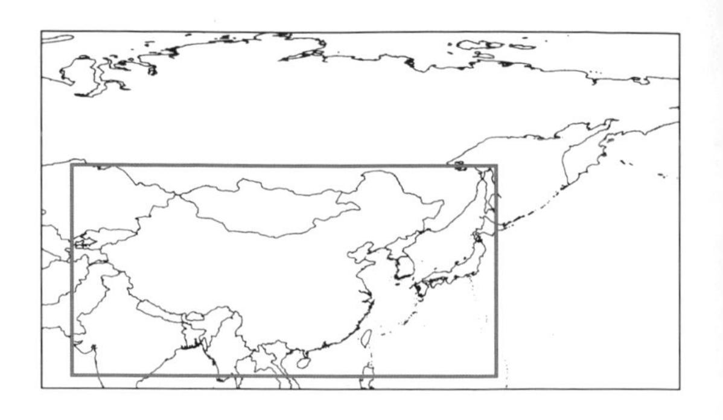

Distribution

There are some ten species in the genus *Taxus* but only five are currently listed under CITES. These species are trees and shrubs native to Afghanistan, Bhutan, China and Taiwan, Democratic Republic of Korea, India, Japan, Malaysia, Myanmar, Nepal, Pakistan, the Philippines, Republic of Korea, Russian Federation and Viet Nam. See Species + for individual species distribution: **http://www.speciesplus.net/#/taxon_concepts?taxonomy=cites_eu&taxon_concept_query=Taxus&geo_entities_ids=&geo_entity_scope=cites&page=1**.

Uses

These species are grown for use as horticultural ornamentals, hedging plants, in particular *Taxus cuspidata*, and bonsai trees. Chemical extracts produced from the leaves and bark are used in anti-cancer drugs. The extracts are also used in traditional Chinese medicines (TCMs) as an anti-diabetic drug.

Trade

Live plants are the dominant product in trade. The CITES Trade Database indicates that the main exporters of artificially propagated live plants are the USA and Republic of Korea with the main importers being China and Italy. The main importer of extracts is Canada (exported by the USA) and China (direct exports from Myanmar). The main re-exporter of extracts is Italy.

The #2 annotation means all timber and parts and derivatives are regulated except seeds, pollen and those products traded as finished products packaged and ready for retail trade. The definition of "finished products packaged and ready for retail trade" can be found in the Interpretation section of the Appendices or Annexes or the CITES Glossary (**http://www.cites.org/eng/resources/terms/glossary.php**) as any "Product, shipped singly or in bulk, requiring no further processing, packaged, labelled for final use or the retail trade in a state fit for being sold to or used by the general public". With live, artificially propagated hybrids and cultivars of *T. cuspidata*, traded in pots or other small containers, they must meet the criteria laid out in the annotation to be considered exempt from regulation. CITES controls also apply to the infraspecific taxa (e.g. cultivated varieties) for all these species, except *T. wallichiana.*

Plantations / artificial propagation

Commercial plantations have been established in China and Viet Nam to supply material for the production of anti-cancer drugs. The crude and finished medicines are traded for the domestic markets but may also be in international trade. Live artificially propagated varieties of these species are in international trade, in particular those of *Taxus cuspidata*.

Taxus cuspidata

Taxus cuspidata ripe seeds

CITES international trade suspensions, export quotas and reservations

There are no current CITES international trade suspensions, export quotas or reservations in place for these species.

There are trade suspensions in place for some of these species including the export of all wild specimens of Appendix I, II and III species from India (see Notification 1999/039 **http://cites.org/sites/default/files/eng/notif/1999/039.shtml**). However, export permits will be issued for cultivated varieties of plant species included in Appendices I and II.

EU Decisions

There are no current EU suspensions or opinions for these species.

See Species + for details **http://www.speciesplus.net/#/taxon_concepts?taxonomy=cites_eu&taxon_concept_query=Taxus&geo_entities_ids=&geo_entity_scope=cites&page=1**

<table>
<tr><th>SCIENTIFIC AND COMMON NAMES</th><th>DATE OF LISTING</th><th>CURRENT LISTING AND ANNOTATION</th></tr>
<tr><td>Scientific names and authors:
Taxus chinensis (Pilg.) Rehder
Taxus cuspidata Sieb. & Zucc.
Taxus fuana Nan Li & R.R.Mill 1997
Taxus sumatrana (Miq.) de Laub.
Taxus wallichiana Zucc.</td><td rowspan="3">Appendix II
T. chinensis: 12/01/2005
T. cuspidata: 12/01/2005
T. fuana: 12/01/2005
T. sumatrana: 12/01/2005
T. wallichiana: 16/02/1995

Annex B
T. chinensis: 22/08/2005
T. cuspidata: 22/08/2005
T. fuana: 22/08/2005
T. sumatrana: 22/08/2005
T. wallichiana: 01/06/1997</td><td rowspan="4">Appendix II
T. chinensis: 13/09/2007
T. cuspidata: 13/09/2007
T. fuana: 13/09/2007
T. sumatrana: 13/09/2007
T. wallichiana: 13/09/2007

Annex B
T. chinensis: 20/12/2014
T. cuspidata: 20/12/2014
T. fuana: 20/12/2014
T. sumatrana: 20/12/2014
T. wallichiana: 20/12/2014

Annotation
#2 All parts and derivatives regulated, except seeds and pollen and finished products packaged and ready for retail trade.
Includes intraspecific of these species, except in the case of T. wallichiana).
Also artificially propagated hybrids and cultivars of T. cuspidata, live, in pots or other small containers, each consignment being accompanied by a label or document stating the name of the taxon or taxa and the text 'artificially propagated', are not subject to the provisions of this Regulation.</td></tr>
<tr><td>Family: Taxaceae</td></tr>
<tr><td>Common names:
Chinese: ximalaya, hongdoushan
English: Himalayan yew (T. wallichiana), Japanese yew (T. cuspidata), Chinese yew (T. chinensis and T. sumatrana)
French: if de l'Himalaya
Italian: teixo
Portugese: tasso
Spanish: tejo del Himalaya</td></tr>
<tr><td colspan="2">CITES Standard Reference: The CoP has adopted a standard reference for this genus (The World Checklist and Bibliography of Conifers, A. Farjon, 2001). See http://www.kew.org/data/checklists.html#conifers and http://herbaria.plants.ox.ac.uk/bol/conifers</td></tr>
</table>

Taxus cuspidata bonsai

Pills – finished product

IDENTIFICATION

Prior to the identification of a sample being carried out, collate all the information you have about it (e.g. country of origin) and ensure you have checked the CITES listing and understand its scope. This may influence the questions you ask:

Identification to species or genus level

What species / genus is it? — no information on the sample is available.
Is it species / genus X? — to confirm that the sample matches a listed species / genus.

Geographical information

Which country is the sample from? — necessary when only specific populations are CITES-listed.
Is the sample from a particular timber concession, region or group of trees? – to confirm if trees or concessions that have been identified for export are in trade.

Source

Is the sample from artificially propagated (cultivated) or wild-sourced timber? – to confirm if certain sources of timber are in trade.

Age

How old is the sample? — to confirm the timber pre-dates implementation of any applicable legislation.

You should also check where the nearest laboratory / institute is that is competent to carry out identification tests; whether verified vouchered samples or comparative reference profiles/databases are in place to aid identification; what the charges are per sample and what is considered one sample (e.g. one paintbrush or a batch of paintbrushes); and how long the test will take.

There are a number of identification techniques available:

ANATOMICAL IDENTIFICATION — where the characteristics (type, distribution and arrangement) of the wood's physical structure are used to identify the wood sample. This method can identify the sample to genus and / or family level and sometimes to species level depending on expertise and availability of validated reference samples. It uses either:

Macroscopic characteristics — visible to the unaided eye or with the help of x10 power and above hand lens. This type of identification can be carried out in the field; OR

Microscopic characteristics — these are too small to be seen with the unaided eye or with a hand lens. A light or electron microscope is required. This is carried out in a laboratory and can include identification of fibres.

When identifying a wood sample using anatomical characteristics it is necessary to understand which plane (referred to as "face", "section" or "surface") you are looking at. The transverse or cross section is the most useful for viewing the wood's structure. If the sample has been heavily processed or is very thinly sliced there may be difficulty in identifying it. For anatomical identification where possible take samples that are at least 2–3cm^3 in size.

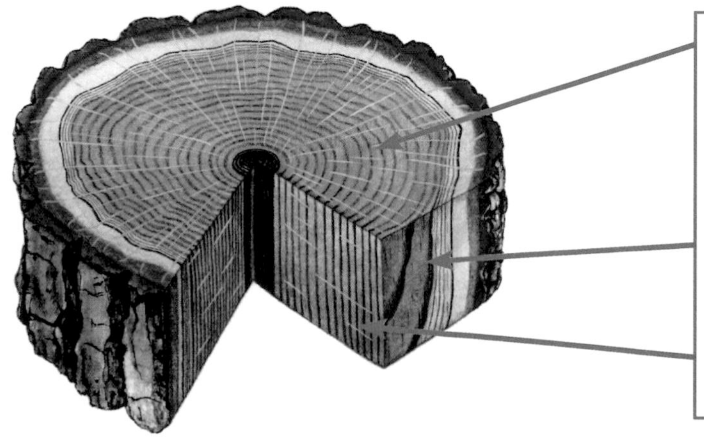

Transverse plane
This plane is sometimes also called a cross-sectional plane or simply a cross section, end or cross grain and often provides the most useful information about the distribution, type and arrangement of the wood vessels.

Tangential plane
A longitudinal plane at right angles to the radius of the stem.

Radial plane
A longitudinal plane along the radius of the stem.

CHEMICAL IDENTIFICATION — there are a number of different techniques available:

Mass spectrometry — analysis of the chemicals synthesised by a plant. This produces a chemical profile which can be matched against reference material/datasets. This method can potentially identify a sample to species/ genus level and differentiate between cultivated versus wild material if sufficient reference material and databases are available.

Radiocarbon dating — used to calculate the age of the sample.

Stable Isotopes — measurement of the ratios of different stable isotopes to produce an isotopic fingerprint that often relates to specific geographic and climatic variables. This method can be used to identify geographic provenance

Near Infrared spectroscopy (NIRS) — this method measures the chemical characteristics in a sample after it has been subjected to near infrared electromagnetic energy. This method can be used to identify different species within a genus and region and individuals of different genera, but is dependent on sufficient validated reference samples being available.

Genetic identification — DNA analysis can usually identify a sample to species level, and may allow the determination of provenance if sufficient comparative reference profiles are available. Methods include:

DNA sequencing (or DNA bar coding) — this method generates a DNA sequence for a specific gene that is typically characteristic of the taxon or geographic region of origin of the sample. The DNA sequence for an unknown sample can be compared against reference data to allow identification. This method can be used to identify a sample to species, genus and family level and occasionally broad geographic origin.

DNA profiling (or DNA fingerprinting) — this method is used to identify genetic differences among biological populations or individuals. DNA profiles can be used to provide a unique identification for individual trees, or to assign a sample to its population of origin. The method can also be used as the basis for exclusion testing in supply chain authentication applications.

OTHER TECHNIQUES

Automated machine vision — a technique still in the prototype stage that uses a handheld device to take an image of the wood's surface allowing comparison against verified reference samples.

Visual aids — a number of CITES manuals and computer-based interactive databases exist to aid identification. They include CITES*wood*ID (see *KEY RESOURCES* for more information) and the CITES Identification Guide – Tropical Timbers (English version at **https://cites.unia.es/cites/file.php/1/files/CAN-CITES_Wood_Guide.pdf** and Chinese version at **http://www.traffic.org/identification/**).

Detector dogs — use of dogs to detect timber and timber products by odour e.g. agarwood in the UK, big-leaf mahogany in Germany **www.traffic.org/non-traffic/non-traffic_pub23.pdf** and **http://www.wwf.de/fileadmin/fm-wwf/Publikationen-PDF/proceedings_of_the_conference_on_wildlife_detector_dogs_budapest_2012.pdf**

Identification of timber seized by UK Border Force

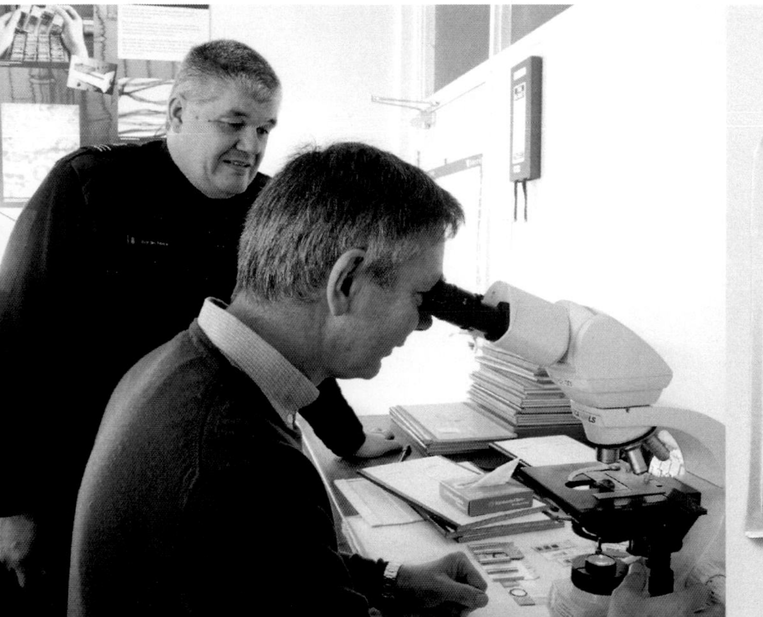

Timber samples from RBG Kew

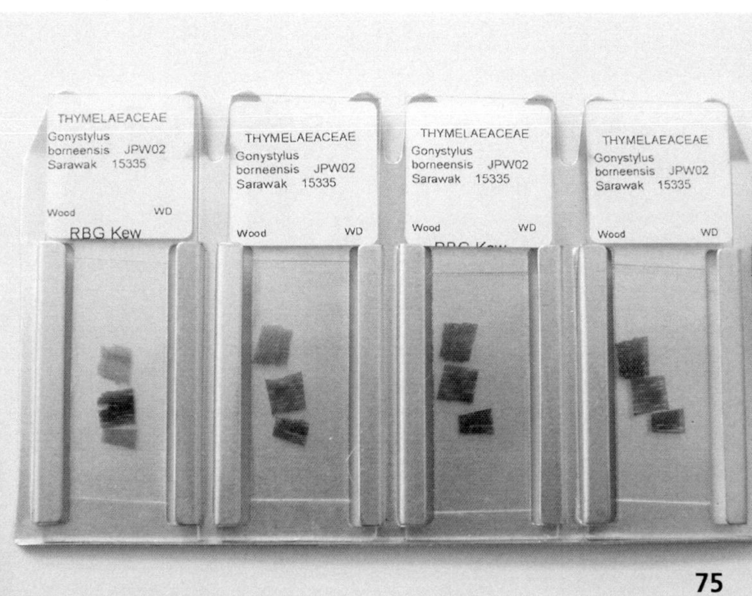

CONTACTS

Royal Botanic Gardens, Kew: **http://www.kew.org/learn/specialist-training/wood-identification-course** and **http://www.kew.org/collections/ecbot/wood/**.
Contact Dr. Peter Gasson **p.gasson@kew.org**

Trace Wildlife Forensics Network: **http://www.tracenetwork.org/wildlife-forensics/what-is-wildlife-dna-forensics/**.
Contact **info@tracenetwork.org**

Thünen Centre of Competence on the Origin of Timber: **https://www.ti.bund.de/en/infrastructure/the-thuenen-centre-of-competence-on-the-origin-of-timber/**.
Contact Dr Gerald Koch **gerald.koch@ti.bund.de<gerald.koch@ti.bund.de**

US Fish and Wildlife Forensic Laboratory: **http://www.fws.gov/lab**
Contact Ed Espinosa **ed_espinoza@fws.gov** and Gabriella Chavarria **gabriela_chavarria@fws.gov**

Global Timber Tracking Network (GTTN): **http://www.globaltimbertrackingnetwork.org/home/**.
Contact **gttn@cgiar.org**

Papers and publications related to timber identification are found in the *KEY RESOURCES* section.

Field identification of seized ramin

Machine vision

Wood collection (Thünen Institute)

TIMBER MEASUREMENT

Timber is traded in many different forms and often the amounts recorded on invoices, permits, etc are in different unit codes (e.g. carvings, sawn wood, etc). To assist you when verifying the quantity invoiced on the shipment's documents matches the quantity recorded on the CITES permit or certificate we recommend you use the following formulas (see *KEY RESOURCES* page for more information on conversion rates)

A number of the units used are specific to certain countries and may not be used in your country e.g. board feet is a unit of volume often used in the USA and Canada.

For assistance, ask the trader whether they use any standard conversion rates or contact your CITES Scientific Authority or your local or national forestry/plant health agency. These conversions should also be made by the importer or exporter so that the total quantity of CITES regulated material recorded on the shipping documents is expressed in the same unit of measurement found on the CITES documentation.

Note: conversion rates are meant for guidance only and are taken from the USDA CITES I-II-III Timber Species Manual (2010) http://www.aphis.usda.gov/import_export/plants/manuals/ports/downloads/cites.pdf

CONVERT

Kilograms (kg) of timber to cubic metres (m^3)

There are 450–700 kg of timber per cubic metre of timber. Use the guide figure of 600 kg

Conversion formula: 1,000 kg / 600 = m^3

→

EXAMPLE

1,000 kg of timber:

1,000 kg divided by 600 = 1.67m^3 of timber

CONVERT

Cubic feet (cubic ft) to cubic metres (m^3)

Square ft x thickness = cubic ft.

1 cubic ft = 0.02832 m^3

Conversion formula:

cubic ft x thickness (ft) x 0.02832 = m^3

→

EXAMPLE

10,000 cubic ft (approx. 1 inch thick):

10,000 sq. feet x 1/12 (12 inches in a foot) x 0.02832 = 23.6 m^3 of timber

CONVERT

Square feet to square metres (m^2)

Convert sq. ft to sq. metres (m^2)
[sq. ft = length (ft) x width (ft)]

Conversion formula: 1 sq. ft x 0.0929 = m^2

→

EXAMPLE

25,000 sq. feet of timber:

25,000 sq. ft x 0.0929 = 2,322.5 m^2 of timber

CONVERT

Square metres (m^2) to cubic metres (m^3)

Convert sq. metres to cubic metres (m^3)

Conversion formula: m^2 x thickness = m^3

EXAMPLE

25,000 m2 of timber (veneer) (0.6mm thick):

25,000 m2 x 0.0006m = 15 m^3 of timber

CONVERT

Volume of a cylinder (inches) to cubic metres (m^3)

Convert volume of cylinder inches to cubic metres.

N.B. p (3.14) x (radius in inches)^2 x (length in inches) x (total number of dowels) = cubic inches.

Conversion formula:
(cubic inches) x (0.0000164) = m^3 of dowel

EXAMPLE

100,000 dowe ls (1/4 inches diameter) x 16 inches in length:

Radius = 1/2 diameter → 1/4 = 0.25 x 1/2 = 0.125

(3.14) x (0.125)^2 x 16 inches x 100,000 = 78,500 cubic inches.

(78,500 cubic inches) x 0.0000164 = 1.287 m^3

CONVERT

Board feet (usually expressed as pie tablares (PT)) to cubic metres (m^3)

There are 424 PT per cubic metre

Conversion formula: 1,000 PT / 424 = m^3

EXAMPLE

1,000 board feet (PT) of timber:

1,000 PT of timber divided by 424 = 2.36 m^3 of timber

CITES DOCUMENTATION

The basic requirements for export of CITES Appendix II listed taxa is that a valid export permit should be issued by a Management Authority, following the advice of the Scientific Authority. Some countries, such as the member States of the European Union, apply stricter domestic legislation and require import permits in addition to the export permit.

Information on CITES requirements can be found at:

http://www.cites.org/eng/disc/how.php and http://www.cites.org/eng/res/12/12-03R16.php

Information on European Union Wildlfe Trade Regulations implementing CITES can be found at:

http://ec.europa.eu/environment/cites/legislation_en.htm

Where an export and import document is required you should check for the following:

1. That the importer and exporter details match.
2. That the country of import and the country of export match.
3. That the issuing Management Authority of export is the same as shown in box 24 of the import permit.
4. That the descriptions of the specimens are the same.
5. That the scientific name is the same on both documents.
6. That the common name is the same on both documents.
   Note: These may vary as some species have more than one common name.
7. If box 16 of the import permit is completed the export permit referred to **MUST** be the one presented.
8. That the CITES Appendix recorded on both documents are the same.
9. The source code may differ, you will need to clarify this with your CITES Management Authority.
10. That the purpose recorded on both permits is the same.
11. Although the quantity may vary, the Import permit must cover (i.e. be equal to or less than) the amount recorded on the export permit.
12. Box 14/15 of the export permit must be completed at the point of export. Failure to do so may mean the permit is not valid at the point of import.

CITES Standard permit/certificate form

Standard permit/certificate form

CONVENTION ON INTERNATIONAL TRADE IN ENDANGERED SPECIES OF WILD FAUNA AND FLORA	PERMIT/CERTIFICATE No. ☐ EXPORT ☐ RE-EXPORT ☐ IMPORT ☐ OTHER:	Original 2. Valid until
3. Importer (name and address)	4. Exporter/re-exporter (name, address and country)	
3a. Country of import	Signature of the applicant	
5. Special conditions *For live animals, this permit or certificate is only valid if the transport conditions conform to the* CITES Guidelines for transport *or, in the case of air transport, to the* IATA Live Animals Regulations	6. Name, address, national seal/stamp and country of Management Authority	
5a. Purpose of the transaction (see reverse)	5b. Security stamp no.	

	7./8. Scientific name (genus and species) and common name of animal or plant	9. Description of specimens, including identifying marks or numbers (age/sex if live)	10. Appendix no. and source (see reverse)	11. Quantity (including unit)	11a. Total exported/Quota
A	7./8.	9.	10.	11.	11a.
	12. Country of origin * Permit no. Date		12a. Country of last re-export Certificate no. Date		12b. No. of the operation ** or date of acquisition ***
B	7./8.	9.	10.	11.	11a.
	12. Country of origin * Permit no. Date		12a. Country of last re-export Certificate no. Date		12b. No. of the operation ** or date of acquisition ***
C	7./8.	9.	10.	11.	11a.
	12. Country of origin * Permit no. Date		12a. Country of last re-export Certificate no. Date		12b. No. of the operation ** or date of acquisition ***
D	7./8.	9.	10.	11.	11a.
	12. Country of origin * Permit no. Date		12a. Country of last re-export Certificate no. Date		12b. No. of the operation ** or date of acquisition ***

* Country in which the specimens were taken from the wild, bred in captivity or artificially propagated (only in case of re-export)
** Only for specimens of Appendix-I species bred in captivity or artificially propagated for commercial purposes
*** For pre-Convention specimens

13. This permit/certificate is issued by:

Place — Date — Security stamp, signature and official seal

14. Export endorsement: 15. Bill of Lading/Air waybill number:

Block	Quantity
A	
B	
C	
D	

Port of export — Date — Signature — Official stamp and title

CITES PERMIT/CERTIFICATE No.

(as amended at CoP14)

CITES Standard permit/certificate form

Instructions and explanations:

These correspond to the black numbers on the permit/certificate form

2. For export permits and re-export certificates, the date of expiry of the document may not be more than 6 months after the date of issuance (1 year for import permits).

3. 3a) **Must** be written in full.

4. The absence of the signature of the applicant may render the permit or certificate invalid.

5. Special conditions i.e. national legislation or conditions placed on the shipment by the issuing Management Authority.

   5a) **Note**: *The full list of abbreviations can be found on the reverse of the Permit.*

   5b) Number of stamp affixed in Box 13.

6. The name, address and country of the issuing MA should already be printed on the form.

7. e.g. *Gonystylus bancanus.*

8. e.g. Ramin.

9. Description of the specimen in trade, including any marks e.g. louvre doors.

10. CITES Appendix listing and source e.g. II, W or Appendix II, Wild.

11. Units should conform to the most recent version of the *Guidelines for the preparation and submission of CITES annual reports.* For timber this should be in m^3 or kg.

    11a). The total no. of specimens exported in the current calendar year (1 Jan – 31 Dec) including those covered by the present permit and the current annual quota for the species (e.g. 500/1000). For timber this should be in m^3 or kg.

12. The country in which the specimens were taken from the wild or artificially propagated.

    12a) Only to be completed in case of re-export of specimens previously re-exported. Country from which the specimens were re-exported before entering the country in which the present document is issued. Enter the no. of the re-export certificate and its date of issuance. If all or part of the information is not known, this should be justified in Box 5.

    12b) No. of the registered artificial propagation operation. "Date of acquisition" only required for pre-Convention specimens.

13. Must be completed by the official who issues the permit. Name must be written in full. Security stamp (if used) must be affixed in this block and must be cancelled by the signature of the issuing official and a stamp or seal.

14. Must be completed by the official who inspects the shipment at the time of export or re-export. Enter the quantities of specimens actually exported or re-exported. Strike out the unused blocks.

15. Bill of Lading/Air way bill number

European Community
Standard permit/certificate form

EUROPEAN COMMUNITY

4 — COPY for the issuing authority — 4

1 Exporter/Re-exporter

PERMIT/CERTIFICATE

- [ ] IMPORT
- [ ] EXPORT
- [ ] RE-EXPORT
- [ ] TRAVELLING EXHIBITION
- [ ] CITES PET OWNERSHIP

No

2. Last day of validity:

CITES — Convention on International Trade in Endangered Species of Wild Fauna and Flora

3. Importer

4. Country of (re)-export

5. Country of import

6. Authorized location for live wild-taken specimens of Annex A species

7. Issuing Management Authority

8. Description of specimens (incl. marks, sex/date of birth for live animals)

9. Net mass (kg)

10. Quantity

11. CITES Appendix

12. EC Annex

13. Source

14. Purpose

15. Country of origin

16. Permit No

17. Date of issue

18. Country of last re-export

19. Certificate No

20. Date of issue

21. Scientific name of species

22. Common name of species

23. Special conditions

This permit/certificate is only valid if live animals are transported in compliance with the CITES Guidelines for the Transport and Preparation for Shipment of Live Wild Animals or, in the case of air transport, the Live Animals Regulations published by the International Air Transport Association (IATA)

24. The (re-)export documentation from the country of (re-)export
- [ ] has been surrendered to the issuing authority
- [ ] has to be surrendered to the border customs office of introduction

25. The [ ] importation [ ] exportation [ ] re-exportation of the goods described above is hereby permitted.

Signature and official stamp:

Name of issuing official:

26. Bill of Lading / Air Waybill Number:

Place and date of issue:

27. For customs purposes only

Signature and official stamp:

Quantity / net mass (kg) actually imported or exported	Number of animals dead on arrival

Customs document

Type:

Number:

Date:

Sample

European Community Standard permit/certificate form

Instructions and explanations:

These correspond to the black numbers on the permit/certificate form

1. Name and address of person, persons, or company exporting the shipment.
2. Permit expiry date.
3. Name and address of person, persons, or company receiving the shipment.
4. Country arriving from e.g. Malaysia.
5. Country entering into e.g. the UK.
6. Location for live wild-taken specimens of Annex A species.
7. Address of Management Authority.
8. Whether the specimen is a live plant, a part or derivative – including any specific labels or markings.
9. Net mass (kg). For timber this should be m^3 or kg.
10. Quantity.
11. CITES Appendix.
12. EC Annex listing.
13. The source of the specimen, e.g. **W** (wild) – this specimen comes from a wild non-regulated environment. The *full list of abbreviations to be used can be found on the reverse of the Permit.*
14. What the specimen(s) is/are to be used for. For example, **T** (Commercial) – may be sold for commercial purposes, **P** (personal use) – only for own use. *The full list of abbreviations can be found on the reverse of the Permit.*
15. Where the specimen was removed from its natural environment
16. If completed on the import permit, the export permit referred to in this box MUST be the permit presented.
17. Date of issue of the permit in Box 16.
18. Country of last re-export if different to box 15 that is the country the specimens were re-exported before entering the country in which the present document was issued. Including the (**19**) Permit number for the movement and (**20**) the Permit date of issue.
19. Permit number for the movement, and
20. The permit date of issue.
21. Latin/scientific name e.g. *Cedrela odorata.*
22. e.g. Ramin.
23. Any special conditions imposed by the exporting country.
24. (Re)export documentation from country of (re)export. NB Not all EU Member States complete box 24.
25. The stamp and signature of the issuing officer and the date of issue of the permit. **Note:** *the signature should cancel through the CITES security stamp if one is used.*
26. Bill of Lading/Air way bill number.
27. Export/ re-export/ import endorsement **MUST** be completed by the officer inspecting the documents on export/ import/ re-export.

European Community

Standard form for non-commercial exchange by scientific institutions

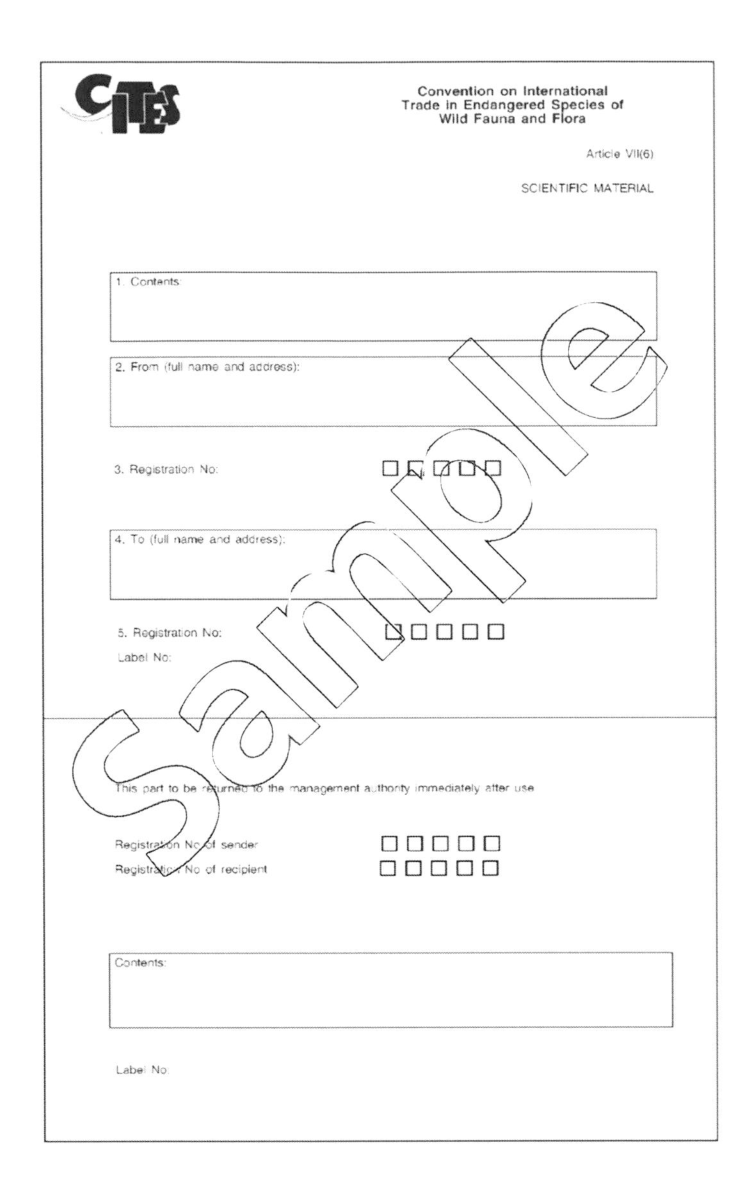
CITES

Convention on International Trade in Endangered Species of Wild Fauna and Flora

Article VII(6)

SCIENTIFIC MATERIAL

1. Contents:

2. From (full name and address):

3. Registration No:

4. To (full name and address):

5. Registration No:

Label No:

This part to be returned to the management authority immediately after use

Registration No of sender

Registration No of recipient

Contents:

Label No:

Sample

There is an exemption from the provisions regulating trade in CITES species that facilitates the loan, donation and exchange of scientific material for non-commercial purposes between scientific institutes. This may include specimens of CITES-listed tree species.

This form is a standard label used by EU registered institutes when they exchange scientific material (see Annex VI **http://eur-lex.europa.eu/legal-content/EN/TXT/PDF/?uri=CELEX:32012R0792&from=EN**). No CITES standard form exists, so non-EU institutes may prepare their own labels. These may differ slightly in their layout.

This label can be used instead of a CITES permit for live or dead plant material, but only between registered institutes. For a list of registered institutes check **http://cites.org/eng/common/reg/e_si.html**

Once registered an institute is given a unique 5-digit registration number. For example, the Royal Botanic Gardens, Kew (UK) is GB 005.

For more information on the criteria an institute must meet to be eligible for this scheme, what information must be displayed on the label and how they can register with their Management Authority see Resolution Conf. 11.15 (Rev. CoP12) (**http://cites.org/eng/res/11/11-15R12.php**), Article 7(4) of Council Regulation 338/97 (**http:/eur-lex.europa.eu/legal-content/EN/TXT/PDF/?uri=CELEX:31997R0338&from=EN**) and Article 52 of Regulation (EC) No 865/2006 (**http://eur-lex.europa.eu/legal-content/EN/TXT/PDF/?uri=CELEX:02006R0865-20120927&from=EN**).

CITES Standard permit/certificate form

Non-commercial cross-border movement of musical instruments

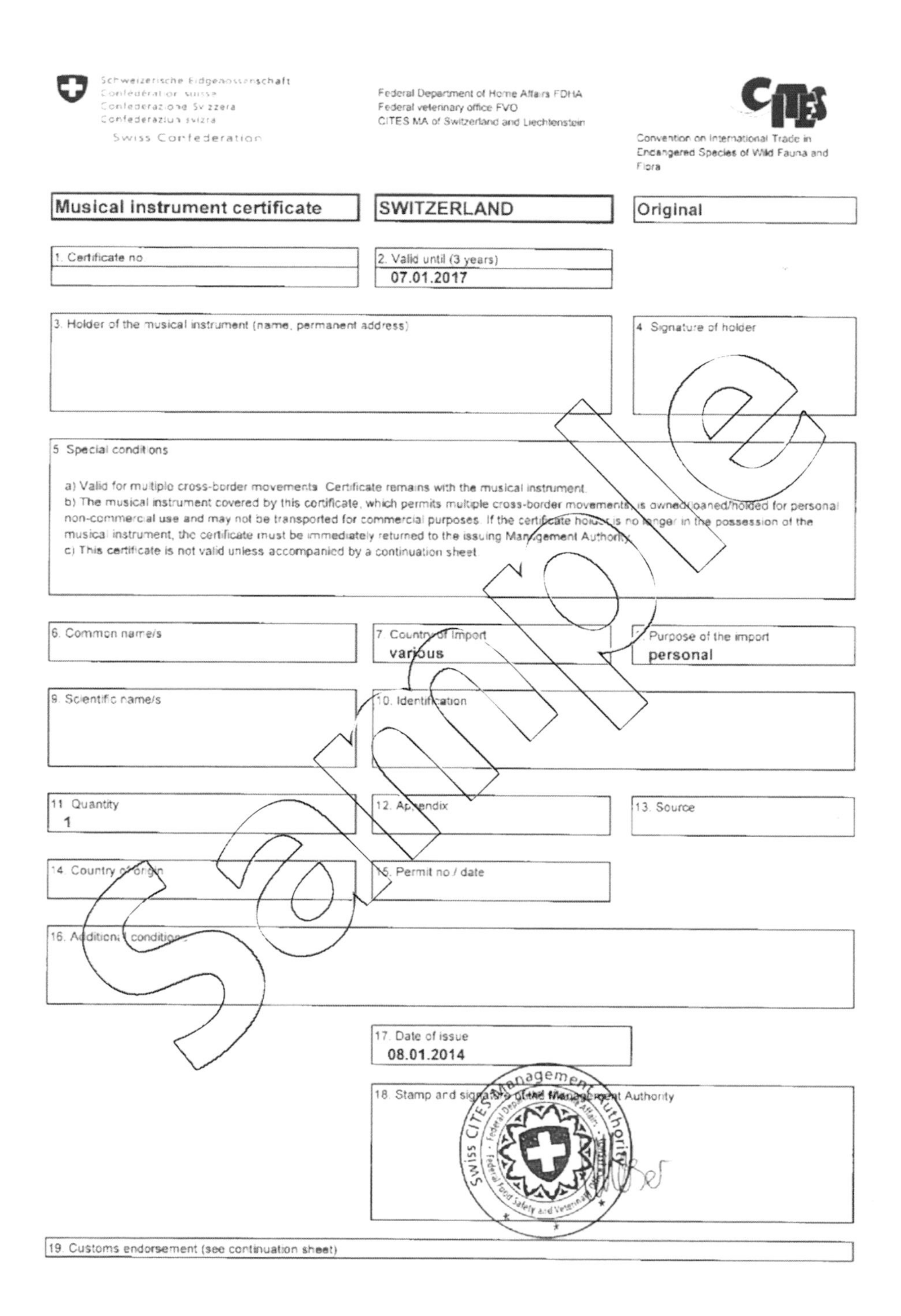

Schweizerische Eidgenossenschaft
Confédération suisse
Confederazione Svizzera
Confederaziun svizra
Swiss Confederation

Federal Department of Home Affairs FDHA
Federal veterinary office FVO
CITES MA of Switzerland and Liechtenstein

CITES
Convention on International Trade in Endangered Species of Wild Fauna and Flora

Musical instrument certificate | SWITZERLAND | Original

1. Certificate no

2. Valid until (3 years)
07.01.2017

3. Holder of the musical instrument (name, permanent address)

4. Signature of holder

5. Special conditions

a) Valid for multiple cross-border movements. Certificate remains with the musical instrument.
b) The musical instrument covered by this certificate, which permits multiple cross-border movements, is owned/loaned/holded for personal non-commercial use and may not be transported for commercial purposes. If the certificate holder is no longer in the possession of the musical instrument, the certificate must be immediately returned to the issuing Management Authority.
c) This certificate is not valid unless accompanied by a continuation sheet.

6. Common name/s

7. Country of import
various

Purpose of the import
personal

9. Scientific name/s

10. Identification

11. Quantity
1

12. Appendix

13. Source

14. Country of origin

15. Permit no / date

16. Additional conditions

17. Date of issue
08.01.2014

18. Stamp and signature of the Management Authority

Swiss CITES Management Authority

19. Customs endorsement (see continuation sheet)

To facilitate the non-commercial cross-border movement of musical instruments derived from CITES species, including CITES-listed tree species, Parties may issue a musical instrument certificate for purposes including, but not limited to, personal use, performance, display or competition.

Parties may use their own certificate (as shown in this example from Switzerland), the CITES standard (re)-export / import permit form (ticking the "Other" box) or they may use a travelling exhibition certificate if an orchestra is moving material by freight. For more information see Resolution Conf. 16.8 (**http://cites.org/eng/res/16/16-08.php**).

European Community Standard permit/certificate form

Non-commercial cross-border movement of musical instruments

EUROPEAN UNION

ORIGINAL

1. Exporter/Re-exporter

PERMIT/CERTIFICATE

- ☐ IMPORT
- ☐ EXPORT
- ☐ RE-EXPORT
- ☐ OTHER:

No

2. Last day of validity:

3. Importer

CITES Convention on International Trade in Endangered Species of Wild Fauna and Flora

4. Country of (re)-export

5. Country of import

6. Authorized location for live specimens of Annex A species

7. Issuing Management Authority

8. Description of specimens (incl. marks, sex/date of birth for live animals)

9. Net mass (kg)

10. Quantity

11. CITES Appendix

12. EU Annex

13. Source

14. Purpose

15. Country of origin

16. Permit No

17. Date of issue

18. Country of last re-export

19. Certificate No

20. Date of issue

21. Scientific name of species

22. Common name of species

23. Special conditions

This permit/certificate is only valid if live animals are transported in compliance with the CITES Guidelines for the Transport and Preparation for Shipment of Live Wild Animals or, in the case of air transport, the Live Animals Regulations published by the International Air Transport Association (IATA)

24. The (re-)export documentation from the country of (re-)export

- ☐ has been surrendered to the issuing authority
- ☐ has to be surrendered to the border customs office of introduction

25. The ☐ importation ☐ exportation ☐ re-exportation of the goods described above is hereby permitted.

Signature and official stamp:

Name of issuing official:

Place and date of issue:

26. Bill of Lading / Air Waybill Number:

27. For customs use only

Quantity / net mass (kg) actually imported or (re)-exported	Number of animals dead on arrival

Signature and official stamp:

Customs document

Type:

Number:

Date:

The EU member states currently use the EU standard import or (re)export permit (ticking the "Other" box) for the movement of musical instruments. See Annex 1 of Commission Implementing Regulation (EU) No 792/2012 (**http://eur-lex.europa.eu/legal-content/EN/TXT/PDF/?uri=CELEX:32012R0792&from=EN**) as amended by Commission Implementing Regulation (EU) 2015/57 (**http://eur-lex.europa.eu/legal-content/EN/TXT/PDF/?uri=CELEX:32015R0057&from=EN**).

For more information see Commission Regulation (EU) 2015/56 (**http://eur-lex.europa.eu/legal-content/EN/TXT/PDF/?uri=CELEX:32015R0056&from=EN**) and section 1.6.7 on musical instrument certificates (**http://ec.europa.eu/environment/cites/info_permits_en.htm#_Toc223858308**).

Also see section 3.6.10 of the Reference Guide to the EU WTR in *KEY RESOURCES* (**http://ec.europa.eu/environment/cites/pdf/referenceguide_en.pdf**).

KEY RESOURCES

TIMBER IDENTIFICATION

CITES*woodID* Version 2008-2 (and its updates). Computer-Aided Identification and Description of CITES Protected Trade Timbers. H. G. Richter, K. Gembruch, G. Koch – enables the user to identify by means of macroscopic characters traded timbers which are regulated by CITES. To use this CD you have to first install the delta program (http://delta-intkey.com/www/programs.htm) and then load the CITESwood database (www.delta-intkey.com/citesw).

CITES Identification Guide – Tropical Woods (2002). Guide to the identification of tropical woods controlled by CITES. Wildlife Enforcement and Intelligence Division, Enforcement Branch, Environment Canada https://cites.unia.es/cites/file.php/1/files/CAN-CITES_Wood_Guide.pdf

Commercial timbers: descriptions, illustrations, identification, and information retrieval. H. G. Richter and M. J. Dallwitz (2000 onwards) http://www.biologie.uni-hamburg.de/b-online/wood/english/gongo-ra.htm

Distinguishing wild from cultivated agarwood (*Aguilaria* spp.) using direct analysis in real time and time-of-flight mass spectrometry http://www.cites.org/sites/default/files/eng/com/pc/21/E-PC21%20Inf.%205.pdf

Evaluating agarwood products for 2-(2-phenylethyl) chromosones using direct analysis in real time and time-of-flight mass spectrometry http://www.cites.org/sites/default/files/eng/com/pc/21/E-PC21%20Inf.%206.pdf

Analysis of select *Dalbergia* and trade timber using direct analysis in real time and time-of-fight mass spectrometry for CITES enforcement http://www.cites.org/sites/default/files/eng/com/pc/21/E-PC21%20 Inf.%207.pdf

Dalnigrin, a neoflavonoid marker for the identification of Brazilian rosewood (*Dalbergia nigra*) in CITES enforcement. Kite, G.C., Green, P.W., Veitch, N.C., Groves, M.C., Gasson, P.E., Simmonds, M.S. Contact: Geoff Kite g.kite@kew.org http://www.kew.org/discover/news/chemistry-aids-conservation

CITES Virtual College – identification tools and information on implementation of CITES https://cites.unia.es/cites/

TIMBER MEASUREMENT

Information on methodology for developing national volume conversion tables (standing volume & export grade sawnwood) – Conversion Table for Sawn Mahogany (*Swietenia macrophylla*) http://www.cites.org/common/com/pc/17/E-PC17-Inf-03.pdf

ENFORCEMENT

EU-TWIX – The EU-TWIX database assists national enforcement agencies, including CITES Management Authorities and prosecutors, in their task of detecting, analyzing and monitoring illegal activities related to trade in fauna and flora covered by the EU Wildlife Trade Regulations. Contact: Vinciane Sacre vsacre@traffic-europe.com http://www.eutwix.org/

The International Consortium on Combating Wildlife Crime (ICCWC) http://www.cites.org/eng/prog/iccwc.php

CITES INFORMATION

Manuals on implementation of CITES timber trade

United States Department of Agriculture CITES I-II-III Timber Species Manual. Provides the procedures for the enforcement of CITES for Appendix I, II and III tree species http://www.aphis.usda.gov/import_export/plants/manuals/ports/downloads/cites.pdf

CITES & Timber. Ramin. Garrett, L., McGough, H. N., Groves, M. and Clarke, G. (2010). Royal Botanic Gardens Kew, UK http://www.kew.org/science-conservation/research-data/science-directory/projects/cites-checklists-and-user-guides

CITES TIMBER TRADE

ITTO-CITES programme on trees species **http://www.cites.org/eng/prog/itto.php**

Ferriss, S., (2014): **An Analysis of Trade in Five CITES-listed Taxa**, London: Chatham House/TRAFFIC**http://www.chathamhouse.org/sites/files/chathamhouse/field/field_document/20140507AnalysisTradeCITESTaxaFerriss.pdf**

CITES timber profiles – Big-leaf mahogany (*Swietenia macrophylla*) **http://www.cites.org/eng/prog/mwg.php**

CITES timber profile – African cherry (*Prunus africana*) **http://www.cites.org/eng/prog/african_cherry.php**

EU INFORMATION

EU implementation of CITES

Reference guide to the EU Wildlife Trade Regulations (EU WTR)
http://ec.europa.eu/environment/cites/pdf/referenceguide_en.pdf

Guide on the differences between CITES and the EU WTR
http://ec.europa.eu/environment/cites/pdf/differences_b_eu_and_cites.pdf

Guide to the opinions of the EU Scientific Review Group (SRG)
http://ec.europa.eu/environment/cites/pdf/srg/def_srg_opinions.pdf)

EU timber trade

Affre, A., Kathe, W. and Raymakers, C. (2004). **Looking Under the Veneer. Implementation Manual on EU Timber Trade Control: Focus on CITES-Listed Trees** by TRAFFIC Europe. Report to the European Commission, Brussels. www.traffic.org/forestry-reports/traffic_pub_forestry9.pdf

Taylor, V., Kecse-Nagy, K. and Osborn, T. (2012). **Trade in *Dalbergia nigra* and the European Union**. Report prepared for the European Commission.

http://www.google.co.uk/
url?sa=t&rct=j&q=&esrc=s&source=web&cd=3&sqi=2&ved=0CDIQFjAC&url=http%3A%2F%2Fec.europa.
eu%2Fenvironment%2Fcites%2Fpdf%2FDalbergia%2520Report_FIN%252020%252012%25202012.
pdf&ei=RqD1VLKcEtXZasasgtAD&usg=AFQjCNFBw37q0v1ec7EFtROqgP1qJRtyYA

CONTACTS

CITES Secretariat. www.cites.org and http://www.cites.org/eng/disc/sec/staff.php

EU Commission. Contact: The European Commission's DG-Environment (Wildlife trade / CITES) env-cites@ec.europa.eu

Environmental Investigation Agency (EIA). http://eia-international.org/

European Union (EU) national agencies concerned with CITES and wildlife trade http://ec.europa.eu/environment/cites/links_national_en.htm

EU-TWIX. Contact: Vinciane Sacré vinciane.sacre@traffic.org

Illegal Logging Portal. http://www.illegal-logging.info/

Interpol – Project LEAF. http://www.interpol.int/Crime-areas/Environmental-crime/Projects/Project-Leaf

IUCN Red List. http://www.iucnredlist.org/

Royal Botanic Gardens, Kew. Contact: Rose Simpson r.simpson@kew.org and Dr.Pete Gasson p.gasson@kew.org

Swedish Environmental Protection Agency. www.naturvardsverket.se www.swedishepa.se

The Plant List. http://www.theplantlist.org

TRACE Wildlife Forensics Network. Contact: Rob Ogden rob.ogden@tracenetwork.org http://www.tracenetwork.org/

TRAFFIC International. Contact: Stephanie Pendry Stephanie.pendry@traffic.org

UK Border Force (CITES Team). Contact: Guy Clarke guy.clarke@hmrc.gsi.gov.uk

UNEP-WCMC (Species +) http://www.unep-wcmc.org/ and http://www.unep-wcmc.org/featured-projects/speciesplus

This guide is available digitally at the websites marked in **green**.

PICTURE CREDITS

The authors would like to thank the following individuals and companies for providing pictures for use in this publication:

Introduction: Environmental Investigative Agency (unloading Hongmu furniture)

Abies guatemalensis: Erik Fernando Alvarado Orellana (plantation, Christmas decoration); Victor Grigas (roof shingles – public domain); Madeleine Groves (tool handles); Andrew McRobb, RBG, Kew (port).

Aniba rosaeodora: Paulo Carmo (plantation, seedlings, logs, oil distillation, unfiltered oil); Catherine Rutherford (essential oil).

Aquilaria* and *Gyrinops: Madeleine Groves (seedlings, resinous heartwood, compressed exhausted powder, essential oil; Andrew McRobb, RBG, Kew (wood chips, perfume); Uwe Schippmann (prayer beads, bracelet).

Page 33 Andrew McRobb, RBG, Kew (sawn wood)

Araucaria araucana: Andrew McRobb, RBG, Kew (*A. araucana* tree); Christoper Notley, Chichester College (veneer sheet); Robert Bishop, Kraftinwood (bowl); Mila Zenkova (petrified cone – this file is licensed under the Creative Commons Attribution-Share Alike 3.0 Unported license); XenoVon (seeds – this file is licensed under the Creative Commons Attribution-Share Alike 3.0 Unported license).

Bulnesia sarmientoi: Andrew McRobb, RBG, Kew (firewood in Brazil, utensils); Miles Gilmer, Gilmer Wood Company (logs); Lance Cruse, Border Force (essential oil drum); Madeleine Groves (essential oil); Margaret Rutherford (mosquito coil).

Caesalpinia echinata: RBG, Kew (*Pernambuco* tree); Fauna and Flora International (unfinished violin bow blanks); Peter Beare (finished violin bows).

Cedrela: Terry Pennington (*Cedrela fissilis*); Rainbow Music, Essex (guitars); RBG Kew (sawn wood)

***Dalbergia* (Madagascar):** Madeleine Groves (guitars); RBG, Kew Madagascar Conservation Centre (logs on beach) Samantha Gunasekara, Sri Lankan Customs (seized Sri Lankan logs, rosewood log).

***Dalbergia* (Asia):** Environmental Investigation Agency (EIA – all)

***Dalbergia* (South America):** RBG, Kew (table); Ana Isabel Fiona Ruiz (*D. stevensonii* logs); Madeleine Groves (chess set); Madeleine Groves, RBG, Kew (seized *D. retusa*); Jim Hafferty (gun blank).

Page 33: Madeleine Groves (shipping containers)

Diospyros: RBG, Kew (tree); Jansamurai (ebony ornaments – public domain); Peter Beare (violin with ebony fingerboard; Madeleine Groves (all remaining)

Dipteryx panamensis: Madeleine Groves (boat, tool handles); Claire Davies, University of Birmingham (hockey sticks – this file is licensed under the Creative Commons Attribution-Share Alike 3.0 Unported license); Andrew McRobb, RBG, Kew (port)

Fitzroya cupressoides: Madeleine Groves (boxes); Corporación Nacional Forestal – CONAP (Chile) (remaining).

Fraxinus mandshurica: Madeleine Groves (table, utensils); Christopher Notley, Chichester College (veneeer sheet); Andrew McRobb, RBG, Kew (port).

Gonystylus: Environmental Investigation Agency (logs); RBG, Kew (sawn timber); Christoper Notley, Chichester College (veneer sheet); Lucy Garrett, RBG, Kew (remaining).

Guaiacum: Miles Gilmer, Gilmer Wood Company (pen blanks);Transforesta SA de CV, Mexico (logs, bearings); Madeleine Groves (pulley sheave, alcohol).

Osyris lanceolata: Quentin Luke (species); Tim Pearce, RBG Kew (logs); Transforesta SA de CV, Mexico (sawdust); Catherine Rutherford (essential oil).

Pericopsis elata: Botanic Gardens Conservation International – BGCI (species); CITES Management Authority, Belgium (logs, sawn wood); Madeleine Groves (flooring); Christoper Notley, Chichester College (veneer sheet).

Pilgerodendron uviferum: Paulo Carmo (house singles); Forest Corporación Nacional Forestal (CONAP), Chile (remaining images).

Pinus koraiensis: Aljos Farjon, RBG, Kew (cones); Madeleine Groves (pine nuts); Andrew McRobb, RBG, Kew (plywood); Victor Sound (logs on lorry – this file is licensed under the Creative Commons Attribution-Share Alike 2.0 Generic license); Christine Johnstone (telegraph poles – this file is licensed under the Creative Commons Attribution-Share Alike 2.0 Generic license).

Platymiscium pleiostachyum: Christoper Notley, Chichester College (veneer sheet); Wikipedia (panelling and flooring – these files are licensed under the Creative Commons Attribution-Share Alike 2.0 Generic license); Andrew McRobb, RBG, Kew (port).

Podocarpus neriifolius: Aljos Farjon (species); Madeleine Groves (oars, utensils, boat); RBG, Kew (table).

Podocarpus parlatorei: Stefan Sauzak (species – this file is licensed under the GNU Free Documentation License); Andrew McRobb, RBG Kew (fenceposts, port); Catherine Rutherford (pencils); Madeleine Groves (brushes).

Prunus africana: UK Border Force (medicine); Terry Sunderland, CIFOR (all remaining).

Pterocarpus santalinus: M. Garg (species – this file is licensed under the licensed under the Creative Commons Attribution-Share Alike 2.0 Generic license); Angie Harms (hanko seals – this file is licensed under the licensed under the Creative Commons Attribution-Share Alike 2.0 Generic license); Ryukei (shamisen – this file is licensed under the GNU Free Documentation License, Version 1.2); Madeleine Groves (powder); RBG, Kew (essential (oil); Environmental Investigation Agency (Hongmu chair carving).

Quercus mongolica: Richard Wilford, RBG, Kew (species); Christine Johnstone (telegraph poles – this file is licensed under the Creative Commons Attribution-Share Alike 2.0 Generic license); Environmental Investigation Agency (logs); Madeleine Groves (tool handles); Gerald Prins (barrels – this file is licensed under the Creative Commons Attribution-Share Alike 3.0 Unported license).

Swietenia: Madeleine Groves (candle sticks); Rainbow Music, Essex (guitars); Pat Ford, U.S. Fish and Wildlife Service (all remaining).

Page 71: RBG, Kew (forest)

Taxus: Madeleine Groves (live plants, medicine); Aljos Farjon, RBG, Kew (seeds); Sage Ross (bonsai – this file is licensed under the Creative Commons Attribution-Share Alike 3.0 Unported license).

Identification: Linda Gurr (cross section line drawing); Madeleine Groves (remaining).

Timber measurement: Ismail Parlan, Forest Research Institute – FRIM (tree measurement).

Page 77: Andrew McRobb, RBG, Kew (logging in Sabah).

Page 80: Andrew McRobb, RBG, Kew (guitar blank).

Front cover: Andrew McRobb, RBG, Kew (forest in Sabah).

Back cover: Transforester SA de CV, Mexico (*Lignum vitae* logs).

ACKNOWLEDGEMENTS

The authors would like to thank the Swedish Environmental Protection Agency for funding this project and the following individuals and agencies for their assistance in the production of this guide:

Mark Albert (U.S. Fish and Wildlife Service); Jonathan Barzdo (independent consultant); Chris Beard (Kew Publishing); Paulo Carmo, (Instituto da Conservação da Natureza e das Florestas, Portugal); Gina Fullerlove (Kew Publishing); Guy Clarke (UK Border Force); Lance Cruse (UK Border Force); Satu Glaser (UNEP-WCMC); Debra Harrison (UK Border Force); Noel McGough (independent consultant); Ursula Moser (Federal Food Safety and Veterinary Office FSVO, Switzerland); Dr Rob Ogden (Programme Director, TRACE Wildlife Forensics Network); Peter Orn (Swedish Environmental Protection Agency); Georgie Smith (Kew Publishing); Anne St John (U.S. Fish and Wildlife Service); Uwe Schippmann (German Federal Agency for Nature Conservation – Bundesamt für Naturschutz – BfN); Terry Sunderland (Center for International Forestry Research – CIFOR); Valentina Vaglica (independent consultant); and Lydia White (Kew Publishing).